Internet of Things

Technology, Communications and Computing

The series Internet Of Things - Technologies, Communications and Computing publishes new developments and advances in the various areas of the different facets of the Internet of Things.

The intent is to cover technology (smart devices, wireless sensors, systems), communications (networks and protocols) and computing (theory, middleware and applications) of the Internet of Things, as embedded in the fields of engineering, computer science, life sciences, as well as the methodologies behind them. The series contains monographs, lecture notes and edited volumes in the Internet of Things research and development area, spanning the areas of wireless sensor networks, autonomic networking, network protocol, agent-based computing, artificial intelligence, selforganizing systems, multi-sensor data fusion, smart objects, and hybrid intelligent systems.

More information about this series at http://www.springer.com/series/11636

Roberto Minerva · Noel Crespi

Networks and New Services: A Complete Story

Springer

Roberto Minerva
Telecom Italia
Turin
Italy

Noel Crespi
Institut Telecom
Telecom SudParis
Evry Cedex
France

ISSN 2199-1073 ISSN 2199-1081 (electronic)
Internet of Things
ISBN 978-3-319-33993-1 ISBN 978-3-319-33995-5 (eBook)
DOI 10.1007/978-3-319-33995-5

Library of Congress Control Number: 2016948259

Printed on acid-free paper

This Springer imprint is published by Springer Nature
The registered company is Springer International Publishing AG
The registered company address is: Gewerbestrasse 11, 6330 Cham, Switzerland

Preface

Communication environments of the future will be fundamentally different than those in use today. One basic assumption underlies all of the reasoning of this book—software is acquiring a definitive role in almost every sector of virtually every industry, including communications. In particular, the integration of new ICT capabilities will determine and shape the evolution of communication systems toward a new smart environment supporting and enriching the human need to "communicate". This integration will undoubtedly shake the traditional foundations of the telecoms industry.

The main focus of the book is to determine whether the communications industry is undergoing a structural transformation or a consolidation of the current technologies, systems, and business models based on connectivity. Its basic statement is that there is a technological paradigm shift under way called softwarization. The possibility that the ICT ecosystem is moving toward the adoption of new technologies that could enable new scenarios and application domains backed up by networking (e.g., smart environments) is thoroughly analyzed.

This book distinguishes itself by its use of a user-centric approach, i.e., an analysis of the technological evolution as driven by the need to satisfy the requirements of people (customers and users). Adopting the user's viewpoint offers a totally different perspective of the technological evolution and can stimulate new and more compelling choices in the service and network frameworks. Specifically, the book shows how different technologies can be integrated with a user-centric approach and how that process offers the possibility of creating a new kind of network enabling new, appealing scenarios.

Covering a range of pertinent subjects, this book presents an analysis of the major technological trends that have (and that will have) an impact on the telecommunications ecosystem, such as software-defined networking, virtualization, and cloud computing. These technologies can be combined in different ways to support "traditional" or more advanced technological infrastructures. Specific business considerations related to technological choices are put forward, followed by a set of possible scenarios characterized by the prevalence of specific technologies and their related business models.

Turin, Italy Roberto Minerva
Evry, France Noel Crespi

Acknowledgments

Behind any book there are many people and even more stories. People's goals and their stories intertwine in many peculiar ways, the story of this book is no different. Actually, its seeds were initially sown as a set of papers and documents whose goal was to convince and then to prove how wrong the telecom operators' mainstream outlook was. Essentially, they did not listen at all. In fact, they took us (the authors) as visionaries, and visionaries who were offtrack at that. Little by little some concepts of the book took shape and grew. They fit together rather nicely, and some people inside the ill-famed R&D organizations of a few Telecom Operators gave us a bit of credit. Here we have to thank those people who shared their insights and suggestions: Antonio Manzalini and Corrado Moiso. Discussions with them were fruitful and rich with new hints and inspiration for further research.

During that same period, an enlightened CEO and his Strategy Vice President found our outlook interesting and listened to our story. One result of that meeting can be summarized in a single sentence: there is a need for a book about new services and new networks. We wish to thank them even if we must keep them anonymous. But we can publicly thank Roberto Saracco, at the time in Telecom Italia, for supporting the work and giving us the freedom to say how things stand in this ever-changing telecommunications world. We are honored to have his preface to our book.

This process made it clear to us that the work we were doing on my (Roberto's) doctoral thesis could have a general value beyond the normal scientific audience, and so we decided to make it into a book. In the meantime, the scientific suggestions and criticisms (because we needed to have qualified people play the role of devil's advocate) of Prof. Tiziana Margaria, Prof. Djamal Zeghlache, and subsequently Prof. Stéphane Frenot and especially Prof. Yvon Kermarrec were fundamental for strengthening the approach and the results. On a more business and deployment perspective, Max Michel's input made the material more consistent, including from the perspective of industry. We gladly thank them all for their help and support. We avoided the criticism of Prof. Thomas Magedanz, which would have surely made the book better, but some friendships are too close to allow all-out technical arguments. So much for the academic and industrial basis and now the fun part.

Have you ever written a book? Forget about being a scientist and a professional spending days and nights capturing difficult concepts and streamlining them into a meaningful sequence of hypothesis and proofs. That was the easy part and actually came out quite naturally and quickly. The tough part was yet to come: consistency checks throughout the book! These activities took a considerable amount of our time and involved a number of other people (not to mention the support of our families). We want to especially thank the people at Springer, in particular Annelies Kersbergen and Maria Bellantone who helped us tremendously, Yasir Saleem and Ehsan Ahvar for their comments and thoughtful review, as well as Miki Nozawa for her ideas and support. And a very special thanks to Rebecca Copeland for her sharp eyes, insightful comments, and the value she brought to this book.

The book is based on a user-centric approach and so it is important to also thank our potential readers that have been in our minds for the entire writing and updating (and indexing) process. We hope that our readers will enjoy this text and recognize (at least some of) the efforts made to offer an engaging book on a very timely topic.

Thank you!

Roberto Minerva
Noel Crespi

Contents

Abbreviation

AI	Artificial Intelligence
API	Application Programming Interfaces
ARM	Advanced RISK Machine
AS	Autonomous System
B2B2C	Business to Business to Customer
CAP	Consistency, Availability and Partition tolerance
CDN	Content Delivery Networks
CLI	Command Line Interface
C–S	Client–Server Systems
CSCF	Call Session Control Function
CSCF	Call State Control Functions
DDOS	Distributed Denial of Service
DHT	Distributed Hash Table
DIY	Do It Yourself
DSLAM	Digital Subscriber Line Access Multiplexer
ETSI	European Telecommunications Standards Institute
FSM	Finite State Machines
GGSN	Gateway GPRS Support Nodes
GPRS	General Packet Radio Service
GSM	Global System for Mobile Communications
HTTP	Hypertext Transfer Protocol
I/O	Input/Output
IaaS	Infrastructure as a Service
ICT	Information and Communication Technologies
IMS	IP Multimedia Subsystem
IoT	Internet of Things
IP	Internet Protocol
IT	Information Technology
KP	Knowledge Plane
LTE	Long-Term Evolution

M2M	Machine to Machine
MANET	Mobile Ad Hoc Network
MMOG	Massively Multiplayer Online Games
MP	Message Passing
NFV	Network Function Virtualization
NGN	Next-Generation Networks
NI	Network Intelligence
NIST	National Institute of Standards and Technology
OAM	Orbital Anngular Momentum
OCCI	Open Cloud Computing Interface
OS	Operating System
P2P	Peer to Peer
PaaS	Platform as a Service
PSTM	Public Switched Telephone Network
QoE	Quality of Experience
QoS	Quality of Service
RCS	Rich Communication Suite
REST	Representational State Transfer
ROI	Return on Investment
RPC	Remote Procedure Call
SaaS	Software as a Service
SDN	Software-Defined Networking
SDP	Service Delivery Platforms
SGSN	Serving GPRS Support Nodes
SIP	Session Initiation Protocol
SLA	Service Language Agreements
SME	Small and Medium Enterprise
SOA	Service-Oriented Architectures
SOAP	Simple Object Access Protocol
SPMD	Single Program Multiple Data
TINA	Telecommunication Information Networking Architecture
UDDI	Universal Description Discovery and Integration
VM	Virtual Machines
VoD	Video On Demand
VoIP	Voice over IP
VSN	Virtual Service Network
W3C	World Wide Web Consortium
WSDL	Web Services Description Language
XML	Extensible Markup Language

List of Figures

List of Tables

Chapter 1
Introduction

This chapter outlines the context, motivation, contributions, and organization of this book. This book aims to discover the right architecture for the emerging types of services and the constantly evolving technologies that are rewriting the whole landscape. The main focal question is whether the telecoms industry is undergoing a structural transformation or a consolidation towards smart environments. The paradigm shift is triggered by three major tectonic plates: the evolution of essential technologies; the metamorphosis of the ecosystems and the value chains; and the advent of new classes of services.

The book explores the main factors of this transformation process: The control paradigms for the provision of services; Networks and service architecture technologies for building viable and effective solutions, with integration of processing, storage, communications and sensing capabilities; New classes of services, e.g. from scientific fields (biology, neuroscience) influencing the new smart environments and challenging the way services are conceived; The ecosystem for new business models, new relationships between actors, new role; new scenarios that range from no change to the most disruptive; and architectural service propositions that consider the assets and the opportunity for Telcos.

1.1 Context

In the digital world, services are a technological and business discriminant; in fact, service providers differentiate themselves by their ability to provide customers a relevant set of easy to use functionalities. Services should be considered from

R. Minerva and N. Crespi, *Networks and New Services: A Complete Story*, Internet of Things, DOI 10.1007/978-3-319-33995-5_1

several perspectives: in terms of functions and interfaces offered to the final users, in terms of supporting architectures and systems, and in terms of proposed business models and related ecosystems. Currently, customers are generally seen as the final users of services, but a number of new trends are changing this belief. In fact, users are now active actors of the service ecosystem. They not only "consume" services and determine the success of commercial offerings, they also contribute to their development and testing; and in some cases, provide the execution platform for their deployment as well.

This evolution is profoundly changing the rules and the foundations on top of which services have been developed and offered thus far. These changes have technological and business impacts along with social and economic consequences.

This transformation is occurring in parallel with another makeover: the traditional world of telecommunications has been drastically changed by the advent of the Internet in terms of new technologies, and new approaches to services, as well as new ecosystems and business opportunities. The intertwining of changes at the service and at the network levels has yet to be fully understood. For instance, the equation that a fall in revenue from communication services can be compensated by major revenues generated by new offerings is highly questionable, and is one of the points dismantled by this work. One of the major changes in this traditional world is due to the "interpretation" that the telecom industry has given to the move from circuits to packets, and the definition and adoption of "all-IP" networks. This passage was mainly viewed as an opportunistic move toward cheaper, more open, more programmable, and more aligned to the Internet infrastructures. To date, the end-to-end principle has been considered in a very dogmatic way, with some advocating for the BellHeads (i.e., the traditionalists and old fashioned operators using IP (Internet Protcol) technologies to refurbish their network) and others for the NetHeads (Isenberg 1998). In reality, this conflict is more intricate and difficult to grasp; Minerva and Moiso (2004) discuss the issues emerging from this argument.

The telecom industry is exposing the need for radical adjustments: consolidated business models and ecosystems as well as architectural solutions and technologies are falling apart. Some historically major companies no longer exist (e.g., Nortel Network), once fierce competitors are now merged and integrated (Alcatel and Lucent, Nokia and Siemens), newcomers emerge and are capable of rapidly obtaining large market share (Huawei, ZTE). Some paradoxes hare arisen, pinpointing the contradictions of this industry. These can be seen as symptoms of consolidation, or signals of even bigger transformation to come. This "consolidation versus structural change" is one of the issues that must be understood in order to properly position the architectural challenges in service architectures and in the telecom industry, in general. Paraphrasing (Kuhn 1996), this could be seen as a period rich in anomalies and paradoxes that are leading to new technical results (that are not necessarily 'better' per se, but which offer promising solutions for the future): in other words, this could be a paradigm shift. As stated in Kuhn (1996):

> The transition from a paradigm in crisis to a new one from which a new tradition of normal science can emerge is far from a cumulative process, one achieved by an articulation or extension of the old paradigm. Rather it is a reconstruction of the field from new fundamentals, a reconstruction that changes some of the field's most elementary theoretical generalizations as well as many of its paradigm methods and applications. During the transition period there will be a large but never complete overlap between the problems that can be solved by the old and by the new paradigm. But there will also be a decisive difference in the modes of solution. When the transition is complete, the profession will have changed its view of the field, its methods, and its goals.
>
> Thomas Kuhn

This seems to be the case of operators. Their goal is to reuse assessed and consolidated technologies to compete with newcomers. They are competing in two different technological fields: in the network and in the service architectures. These architectures follow a release logic: anytime a new architecture is standardized, implemented and released, it is built on top of previously achieved results. Anytime new technologies are considered and introduced in the framework, they are modified in such a way that they fit into this continuity line and can be integrated into the next release provided by Vendors (and they lose their effectiveness). This is clear in the evolution of NGN (Next Generation Networks) and their standards: they are continuously reworking and improving the concept of the Intelligent Network.

1.2 Motivation

In order to determine whether the transformation phase of the telecom industry is leading to a paradigm shift, it is important to consider the current technologies used by the operators and those used by their major competitors and to see what are the respective advantages and capabilities. In addition, the evolutionary path of strongly related technologies in the Information and Communication Technologies (ICT) area can give a detailed insight of the capabilities and possibilities that will be available in the next ten years.

In this timeframe, the pervasiveness of "smart objects," i.e., objects capable of interacting with the external world in order to create an (physical) environment appropriate for providing personalized services to humans and systems, will be very well consolidated.

A high-level definition of service is used here: a service is a time perishable, intangible experience performed for a customer acting in the role of a coproducer (Fitzsimmons and Fitzsimmons 2003). In other words, a generic service refers to processes, functions, performances, or experiences that one person or organization does for the benefit of another. When the service is provided by means of a software system, the adopted definition is: "(Web) services provide a standard means of interoperating between different software applications, running on a variety of platforms and/or frameworks" (Booth et al. 2004).

The following definition can be considered: "A Service System is a complex system, i.e., a system in which the parts interact in a nonlinear way. As such, a service system is not just the sum of its parts, but through complex interactions, the parts create a system whose behavior is difficult to predict and model. In many cases, the main source of complexity in a service is its people, whether those at the client, those at the provider, or those at other organizations" (Spohrer et al. 2007).

Nowadays the most used control paradigm is the well-known client–server model. Other two relevant models are the NI (Network Intelligence) and the P2P (Peer-to-Peer) paradigms. These paradigms are used to provide a large number of successful services. Other models are currently emerging, especially for supporting new classes of services: e.g., the PubSub model. The type of distributed infrastructure and the control paradigms to be used to provide services are a major concern of and a differentiator between competitors. It is important **to understand what are the technologies and the control paradigms to use in order to implement services in an effective way, i.e., providing a rich set of functionalities, supporting extensibility, and allowing different levels of programmability**. This should also be considered from the point of view of the **expressive power of the solutions** emphasizing the importance of identifying to what extent a control paradigm can fulfill richness, extensibility, and simplicity requirements to implement services and functionalities in highly distributed and pervasive systems.

Being able to understand these technologies at this level leads to another major issue: **what is the right architecture for (new) classes of services**? A large part of the telecom industry has been striving to promote horizontal solutions, i.e., general purpose platforms over imposed to the network infrastructure that are to be used to provide any sort of service. Many attempts to provide a generic and full horizontal architecture have failed. The TINA (Telecommunication Information Networking Architecture) architecture (Berndt and Minerva 1995) had this ambitious goal but it could not deliver. Today, the "web as a platform" trend is revealing that web technologies and the related infrastructure are a viable way to propose a large number of vertical services. **This book aims to identify the fittest control paradigms to be used to support the infrastructure of the future**. In addition, **it will try to give a glimpse of the possible classes of services that are technically and economically viable for the future**.

It is important to emphasize that in the generic definition of a service, the user has a proactive role and is directly involved in the service provision. This aspect is not obvious in the definition of service for the ICT industry, where in fact, the user is not properly identified as a proactive actor of the provision of services. From this standpoint, another characterizing aspect of this work is the "User Centric approach." Generally the user-centric approach is utilized to identify the requirements of users and to improve the means to satisfy them in building up a product or a software solution (Eriksson et al. 2005). In this document, the "User Centric approach" goes further and embraces the definition of service and service system: it emphasizes the possibilities and the capabilities that a user could make available directly in order to cooperate in the creation, development, and offering of services and platforms. In other words, the ICT capabilities are such that users can create their own infrastructure for using services and applications independently of other actors. Why is this important? Many times, the rights and the capabilities of the user are disregarded and neglected in favor of the (economic) interests of specific actors of an ecosystem. As explained in Clark et al. (2005), many Internet solutions are implemented, not to achieve the best technical result in the interest of users, but merely to exploit and consolidate the techno-economic advantages of a particular actor. Many services could be offered in very different and more effective ways if the user was really posed at the center of the service design. **One objective of this work is to approach the service architecture definition with a user-centric approach and hence to identify the situations that affect users the most**.

In order to determine how services and service architectures will evolve, it is important to identify roadmaps for future technologies that will have an impact. One of the major questions is: will the foreseen evolution of technologies bring about a different way to provide services? How will these new technologies reshape the network and service frameworks? In order to answer these questions, **this book will extensively discuss the impact of new technologies on service and network architectures**. A byproduct of this investigation is an attempt to **determine (at a very general level) that what will be the value of network and services in the future**. In fact, many disruptive technologies are emerging (e.g., cognitive radio, software-defined networks) that can drastically redefine the value of the network asset. **One hypothesis that will be investigated is how the combination of general purpose computing and software will impact the telecom infrastructure**. This possibility is well represented by the concept defined in Andreessen (2011) that "**software will eat everything**."

1.3 Focus of the Book

The future communications environment will be different than the current one. The integration of different capabilities will determine and shape the evolution toward a new smart environment supporting and enriching the human need to "communicate." This smart environment will put together communications, processing, storage, and sensing/actuating capabilities. This integration will likely shake the traditional foundations of the telecom industry. Figure 1.1 depicts the convergence of these four areas (in the Enabling technology oval) into a powerful setting that enables smart environments. The extension of this enabling environment with advanced software technologies will lead to the creation of a new sector, the smart environment.

On top of the basic infrastructure (composed of enabling technologies), the ancillary technologies can bring further value, such as cognitive and artificial intelligence, distributed processing solutions, cloud and grid computing, virtualization, new means to deal with data, which can shape a different intelligent environment for users to be immersed in.

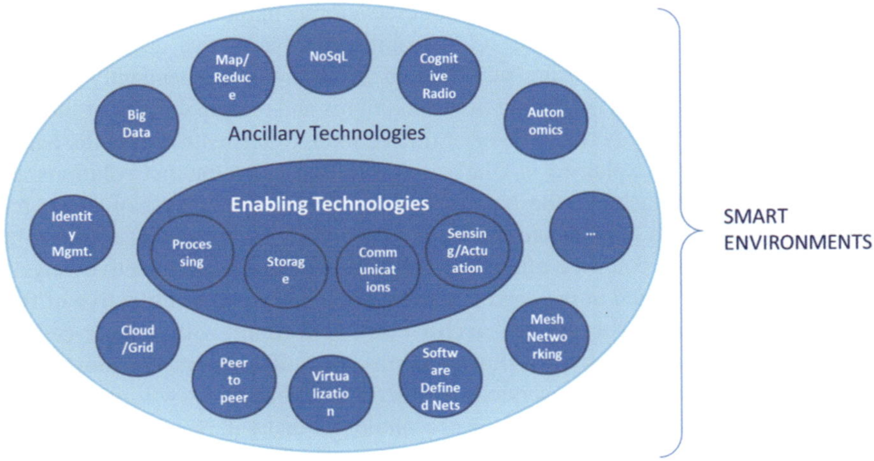

Fig. 1.1 Convergence of technologies will lead to smart environments

The main focus of the book is related to determining **whether the telecom industry is undergoing a structural transformation** or a consolidation toward smart environments. The main assumption is that **there is a paradigm shift underway**. This change is manifest and evident in terms of

- The evolution of essential technologies, i.e., the book identifies and shows the general technological evolution and specific disruptive technologies that could alter the status quo;
- Modifications of the ecosystems, i.e., the book identifies the assets of different actors with major roles in the industry and discuss how they are reacting to this shift. Special attention is paid to how TelCos can respond to this turmoil; and
- The advent of new classes of services and their appeal, i.e., the identification of service opportunities. These are analyzed in terms of the required control paradigms and the related service architectures for supporting these services.

Another major effort is to identify **the possibilities of enforcing an actual and accurate user-centric approach** that truly empowers users and is aligned with their rights and capabilities. This means being able to demonstrate when and how users could have a proactive role in service provisioning and in the construction of related platforms. The universe of discourse of the books could be broadly categorized as shown in Fig. 1.2.

In particular these factors are considered and thoroughly analyzed

- *Control paradigms for the provision of services*: in a highly pervasive and distributed environment, the way objects communicate and interact is of paramount importance. Adopting the wrong interaction paradigm can undermine the entire service offering and hamper the smooth adoption of meaningful services.

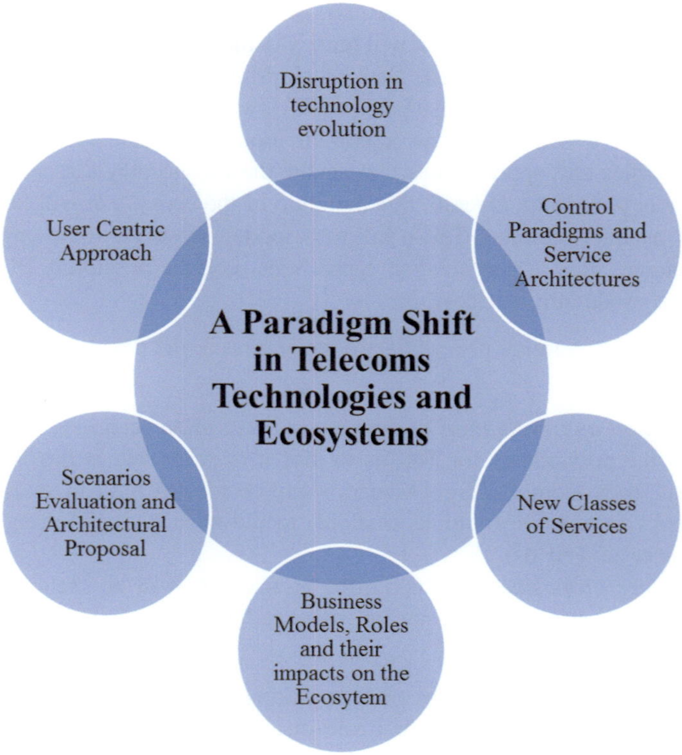

Fig. 1.2 Scope of the analysis

- *Networks and service architecture technologies for building viable and effective solutions*: service and network architectures should be capable of supporting the integration of processing, storage, communications, and sensing capabilities. Many of the current architectures are not designed to allow this integration. There is a real need to reshape, refurbish, and even to throw away current architectures in order to support the new smart environments. Actors that have better architectures will have the edge. This analysis will especially focus on the possibilities for TelCos to offer appropriate and viable platforms to address the future communications environment.
- *New classes of services*: the combination of new technologies with the proper control paradigms will lead to the creation of new classes of services that will be substantially different from the current ones. The most important aspect of this convergence is that services will also arise from other scientific fields (e.g., biology, neuroscience, and physics) and their combination with smart environments will drastically challenge the way services are currently conceived. It will not be the Internet of Things, but every single real-world object will be deeply embedded in the Internet.

- *Mutation of the ecosystem, new business models and roles*: smart environments and the technological evolution will have a tremendous impact on how services and applications are provided. New relationships between actors will emerge and the role of users could change as well. New actors could play relevant roles (e.g., local communities). One important proposition here is to identify and describe roles and opportunities that enable the user to play a proactive role in the new environment. Overall, this mutation of the ecosystem will have a huge impact on the traditional TelCo business model. However, new opportunities will emerge, such as servitization capabilities (i.e., the possibility to transform products/goods into services).

In order to give a future perspective to the approach, the next steps should focus on these aspects

- *Construction and evaluation of future scenarios*: these scenarios will take into account the possibilities for TelCos to play a relevant role in the "smart environment" ecosystem. Different scenarios will be constructed and analyzed, from the most conservative (nothing changes) to the most disruptive (e.g., the disappearance of TelCos).
- *Architecture propositions*: a supporting architecture will be sketched for each viable role as a way to emphasize the assets and the opportunity that a TelCo can put into the field.

There is an underlying consideration supporting all the reasoning of this work—that software is acquiring a principal role in almost every sector of virtually all industries. As clearly put forward by Andreessen (2011), the question is "Why Software will eat everything." If one sentence should be used to describe this book, then "how software will eat the traditional telecommunications industry" is the one to use. This theme will be developed in the book as described in the next section.

1.4 Organization of the Book

Based on the current research trends, this book is organized into three parts.

The first part covers the technology evolution in terms of essential technologies, new network architectures, and new classes of services. Contextually this part also describes the different control paradigms and how they are leveraged to support the solutions. A broad overview of the literature is presented in Chap. 2 and provides a general framework for the technological issues that will be further discussed in the book. This overview is followed in Chap. 3 by an analysis of the possible evolution of enabling technologies that are relevant for the progress of the ICT sector, and new network architectures are discussed in Chap. 4. A perspective on how the network infrastructure may evolve and which functionalities and services could be offered and supported is provided in Chap. 5, along with a comprehensive view on emerging services and their control paradigms with respect to the new network architectures. Services that largely differ from the normal communication capabilities are considered as new possibilities and opportunities for really entering the service age. Finally, an analysis of how the TelCos are approaching this service age is provided in Chap. 6, highlighting the importance of software in contrast to the importance of communication.

Bibliography

Andreessen M (2011) Why software is eating the world. Wall Street Journal Available at http://online.wsj.com/article/SB10001424053111903480904576512250915629460.html. Accessed May 2013

Berndt H, Minerva R (1995) Definition of service architecture. Deliverable. TINA Consortium, Red Bank

Booth D et al (2004) Web services architecture. W3C Working group note, W3C, W3C

Clark DD, Wroclawski J, Sollins KR, Braden R (2005) Tussle in cyberspace: defining tomorrow's internet. IEEE/ACM Trans Netw (TON) 13(3):462–475

Eriksson M, Niitamo VP, Kulkki S (2005) State-of-the-art in utilizing living labs approach to user-centric ICT innovation-a European approach. White Paper. Lulea University of Technology Sweden, Lulea

Fitzsimmons JA, Fitzsimmons MJ (2003) Service management: operations, strategy, and information technology. McGraw-Hill, New York

Isenberg DI (1998) The dawn of the "stupid network". netWorker, pp 24–31

Kuhn TS (1996) The structure of scientific revolutions, 3rd edn. University of Chicago Press, Chicago

Kuhn TS (2004) The death of network intelligence? In: Proceedings of international symposium on services and local access 2004 (ISSLS). Edinburgh

Minerva R, Moiso C (2004) The death of network intelligence? In: Proceedings of international symposium on services and local access (Edinburgh ISSLS)

Spohrer J, Maglio PP, Bailey J, Gruhl D (2007) Steps toward a science of service systems. IEEE Comput Soc 40(1):71–77

Chapter 2
Service Control Paradigms and Network Architectures

Telcos' viewpoint is "Network is King". Webcos' viewpoint is "Service is King"—the web client and its services are the center points. However, it is not easy to define what "service" is: a bunch of resource; collection of tasks; particular interactions, and semantics; or all the above. By exploring the concepts of service, we can see how services and their definitions evolce, and how their gravity center shifts to unfamiliar zones. The service control is shifting from client–server interaction mode to more open but perplexing peer-to-peer mode. This has a knock-on transforming effect on service structure, message channeling, parallel processing, etc.

Service network architecture has already undergone radical changes with the advent of Cloud and distributed platforms. A middleware layer is bringing edge of the network services back into the network, so it is no longer "stupid". The Telecom's roll out of IMS is decoupling service control from the transport network, but it perpetuates the Telecom network-centric approach. Web style services challenge the perimeterization of services—geographical scope and the Quality of Service (QoS) supremacy. The future service control paradigm must cope with multiplicity of service architectures, serving client-server, network intelligence, and peer-to-peer, all at the same time.

2.1 Introduction

Intuitively, services in the context of Information and Communications Technologies (ICT) are a set of discrete and valuable functions provided by a system. Such a system, in the context of communications, will be distributed, i.e., parts and functions of it are not placed in the same location. In addition, a multitude of users will be requesting for the service. The parts of a distributed system as well as the system and the users need to interact and hence some interaction models and mechanisms should be defined in order to well support the provision of the service

R. Minerva and N. Crespi, *Networks and New Services: A Complete Story*,
Internet of Things, DOI 10.1007/978-3-319-33995-5_2

to users. Since services are provided by a software infrastructure, a general architecture of these systems (and their subsystems) needs to be identified in order to facilitate the design and the life cycles of services. ICT and Communications services presume also the ability to exchange information and hence networks are to be used.

Services can be interpreted in different ways depending on the perspective, the Telecommunication Operators (or Telcos) see services as an extension of basic communication means, Web Companies (or WebCos) see services as functionalities that they can provide/sell by means of web capabilities, other IT (Information Technology) actors consider services as software packages that can be provided and sold to users that will use them in order to accomplish a set of tasks or functionalities (e.g., a spreadsheet). It is important to have a common ground for expressing and discussing the meaning of services.

Services are accessed by the users in a very different way and the internal working of services is characterized also by different control means (i.e., the ways in which functionalities are coordinated within the "service system" and they made available to the users). Also from this perspective, the views on services are profoundly different: Telcos see services as sets of functionalities that deeply rely on the network. Actually the network is driving the service provisioning by requesting the provision of functionalities that the control layer of the network is not capable of delivering. WebCos consider services as a set of functions requested in a client—server functions (the WebCos acting as Server) that can be accomplished with a complex and highly distributed back end system. Typically a WebCo is seen as a "server". Other approaches to services, for instance in Peer-to-Peer or overlay models, adopt different interaction and control paradigms on top of a logical platform. The platform is a means to integrate valuable functions provided by many peers or entities and generic functions needed to support a logical view of the system that is delivering the functions themselves. From this standpoint, also a system devoted to deliver services is conceptually and practically very different depending on the stakeholder that is offering the service functionalities and is building and operating a system supporting the delivery of functions. These differences result in distinct service architectures that do emphasize the approach pursued by the specific Actor and do leverage the specific assets of the Stakeholder (e.g., the network, the data center, the logical overlaying of functions).

The relationship between the network and the services is in fact an important aspect to consider and define, in order to properly understand the intricacies of ICT and Communications services. For Telcos, it is the major asset to leverage in providing services (actually services can be seen as "extensions" or enrichments of the major capability offered by the network: the network is providing the basic service, i.e., connectivity, and services rely on it and functionally extend its properties). For WebCos, the network is just a pipe that is providing a single functionality: the best effort connectivity. All the relevant functionalities are to be provided to the edge and in particular by powerful servers. In peer-to-peer systems, the network is logically abstracted and its relevant properties are offered as services to be integrated with other capabilities and functions offered by the peers.

In order to shed a light on these different approaches and their consequences, a set of examples of some services already implemented adhering to different approaches is presented. Some existing analysis, e.g., (Minerva et al. 2011; Minerva 2008a, b, c), address many of the topics of this chapter and could be considered as a sort of guide to these issues. It also considers the interaction and control paradigms for their expressive power, i.e., the capability to represent and implement the services.

2.2 Services

As a starting point for a Service definition, we will consider the World Wide Web Consortium (W3C) definition (Booth et al. 2004) because it is generally applicable to represent the concept of a service: "*a service is an abstract resource that represents a capability of performing tasks that represents a coherent functionality from the point of view of provider entities and requester entities. To be used, a service must be realized by a concrete provider agent*[1]". A service is then a complex concept that involves many actors/agents that offer resources supporting valuable functions. These functions can cooperate in order to achieve the goals of a service. According to W3C, a Service can be represented as in Table 2.1.

Figure 2.1 depicts the complexity of the service definition.

The W3C definition of a service is a generic one, but it contains some elements that pertain to the specific context and interaction model that are underpinning the web services. For instance, even if it is not cited, the underlying interaction model is client-server in which a requester agent (a client) will send messages (requests) to a provider agent (i.e., a server). The service is a resource that is strictly owned by a person or an organization; this approach is straightforward for the web context (based on a client-server interaction paradigm), but not necessarily fits for other interaction paradigms (e.g., the peer-to-peer). The actions performed by a service aim at achieving a goal state (a predetermined situation favorable to the requester and/or the provider) this is not always the case for other systems based on other interaction paradigms.

Also the Telecom related definitions of service are in a way the brain child of a biased vision of the service world, for instance: "*telecommunications service: 1. Any service provided by a telecommunication provider. 2. A specified set of user-information transfer capabilities provided to a group of users by a telecommunications system. Note: The telecommunications service user is responsible for the information content of the message. The telecommunications service provider has the responsibility for the acceptance, transmission, and delivery of the message*" (NTIA 1996). In addition the term "service" in a telecoms world is very frequently

[1]An agent is a program acting on behalf of a person or organization.

Table 2.1 W3C service definition

Property of a web service	Description
• A service is a resource	• A resource is an entity that has a name, may have reasonable representations and which can be owned. The ownership of a resource is critically connected with the right to set policy on it
• A service performs one or more tasks	• A task is an action or combination of actions that is associated with a desired goal state. Performing the task involves executing the actions, and is intended to achieve a particular goal state
• A service has a service description	• A description is a set of documents that describe the interface to service and its semantics
• A service has a service interface	• An interface is the abstract boundary that a service exposes. It defines the types of messages and the message exchange patterns that are involved in interacting with the service, together with any state or condition implied by those messages
• A service has a service semantics	• It is the expected behavior when interacting with the service. The semantics express a contract (not necessarily a legal contract) between the provider entity and the requester entity. It expresses the intended effect on the real world caused by invoking the service. Service semantics may be formally described in a machine-readable form, identified but not formally defined, or informally defined via an "out of band" agreement between the provider entity and the requester entity
• A service has an identifier	• An identifier is a unique unambiguous name for a resource
• A service has one or more service roles in relation to the service's owner	• A service role is an abstract set of tasks, which is identified to be relevant by a person or organization offering a service. Service roles are also associated with particular aspects of messages exchanged with a service
• A service may have one or more policies applied to it	• A policy is a constraint on the behavior of agents or people or organizations
• A service is owned by a person or organization	• An agent has the right and authority to control, utilize and dispose of the service
• A service is provided by a person or organization	*An agent has the right and authority to provide and operate the service*
• A service is realized by a provider agent	• A provider agent is an agent that is capable of and empowered to perform the actions associated with a service on behalf of its owner—the provider entity

interpreted as a "network" or at least as the transport service that a network provides, for instance: "*3G: Third-generation mobile network or service. Generic name for mobile network/service based on the IMT-2000 family of global*

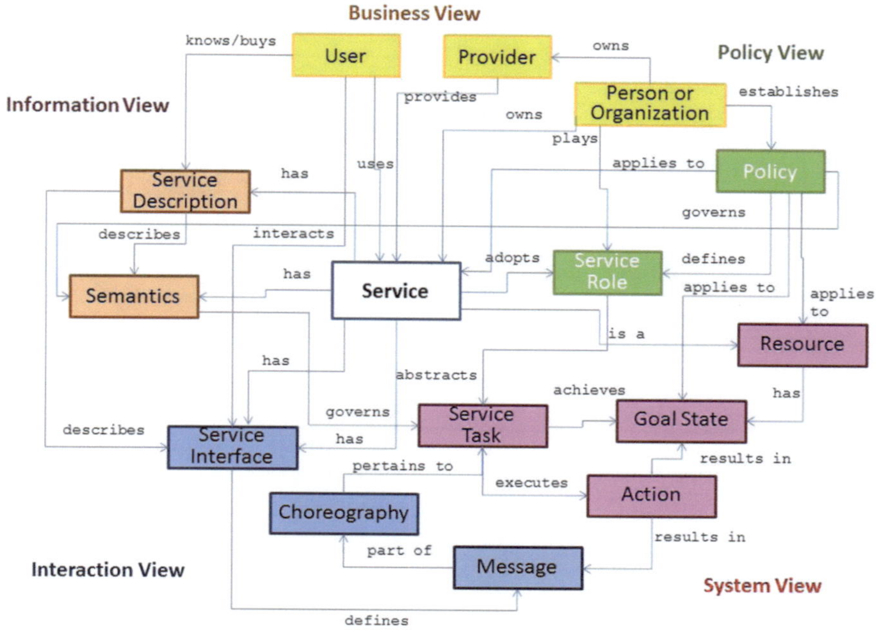

Fig. 2.1 The W3C service definition [(Booth et al. 2004) Fig. 2.8]

standards" (ITU 2011). Often also the literature follows this approach such as (Znaty and Hubaux 1997) that stated:

- "*Telecommunication services is a common name for all services offered by, or over, a telecommunication network*";
- "*Value added services are services that require storage/transformation of data in the network and are often marketed as stand-alone products. Examples of such services are freephone, premium rate, virtual private network and tele-banking. Many value added services can be offered by special service providers connected to the network*";
- "*Teleservices include capabilities for communication between applications. Teleservices are supplied by communication software in the terminals. Telephony, audio, fax, videotex, videotelephony are examples of teleservices*".

All these definitions bring in the concept of a network supporting a major service on top of which "added value" functions and services can be instantiated as represented in Fig. 2.2.

These definitions reflect the vision of services: the network is the main asset and it provides a main service, THE voice service, on top of which several functions or services can be provided. They add value to the unique service at hand.

A more detailed study on service definition that is also considering the integration of different paradigms is given in (Bertin and Crespi 2013): "*a service is*

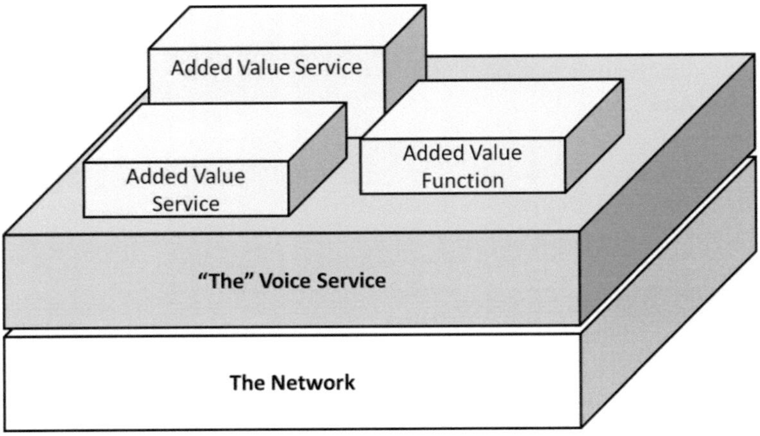

Fig. 2.2 The network, the service and added value functions/services

defined ... as an immaterial business activity made available by a service provider and offering an added value to a user".

Also in the peer-to-peer sector, the service definition strongly depends on the "view" of the system organization for supporting functionalities. For instance (Gradecki 2002) defines a service as "*predefined functionality that can be utilized by peers*" where a peer is "*an application, executing on a computer, that has the ability to communicate with similar peer applications*".

These definitions introduce another issue: the differences between services and applications. Actually services and application are sometimes used as synonymous, so an application can be defined as a set of functions that have meaning and usefulness within a specific domain or a software program. In order to differentiate the two concepts, a service is an application that can serve other applications, i.e., a service is an application that offers reusable functionalities that other applications can use over time.

Simply put, the definition of a service is a complex matter that deserves more than a mere technology approach, as stated in (Jones 2005): "*A service's intention is to undertake certain functions to provide value to the business; its specification isn't just the direct service it provides but also the environment in which it undertakes those functions*". This statement points to the problem that services have also nonfunctional goals and they could be so important to determine the meaning and the context of the service itself. Also the previous definition, for instance, could be seen as biased: it mentions the value for business, but a service could have a value for a community independently from any commercial value and business. A User Centric view on services could also change the perspective. Paraphrasing the previous sentence, it could be stated that "*a service's intention is to undertake certain functions to provide value to a person, a group of persons, a community or a business organization and its representatives; its specification isn't just the direct service it provides but also the environment in which it undertakes those functions*".

Services are becoming a crucial topic for many industries; as a consequence a new area is attracting interest from researchers: the Service Science (2013). It is an interdisciplinary approach to the study, design, and implementation of services systems (Service Science 2013). In this context, a service is even more difficult to define because the definition encompasses business, social, technology, and ecosystems issues. A service system, i.e., a system that can support services is a complex system in which specific arrangements of people and technologies take actions that provide value for others.

As stated in (Cardoso et al. 2009), "*the set of activities involved in the development of service-based solutions in a systematic and disciplined way that span, and take into account, business and technical perspectives can be referred to as service engineering: Service Engineering is an approach that provides a discipline for using models and techniques to guide the understanding, structure, design, implementation, deployment, documentation, operation, maintenance and modification of electronic services* (*e-services*)". They propose a new approach to services: services are tradable and can be supported by Internet systems. This combination leads to an Internet of Services that needs to be supported by appropriate Service Architectures and by the Service Engineering.

Summarizing, services strongly depend on many factors like its goals, foreseen users, the ecosystem in which it is used, the system used to be provided and the supporting technologies and mechanisms for using it.

2.3 Interaction and Control Paradigms

Services considered in this document refer to the functions and applications that can be realized by means of ICT technologies. In this context, a service system can comprise different subsystems each one providing specific functions and executing particular tasks. In addition the system as a whole needs to interact with its users. Actually the way in which a service system interacts with users is a characterizing factor of the system itself. In fact, one first consideration is that the users can be seen as external (Fig. 2.3) to the service system or as part of it (Fig. 2.4).

In this document the way in which parts and components of the whole service system interact and cooperate is seen as the "control paradigm", while the way in which external entities interact with the service system is termed as the "interaction paradigm". In the case of control paradigm, the emphasis is on the architecture of the service system, while in the interaction paradigm, the focus is on the way users are accessing the system functions. This separation is not a hard one, in fact interaction and control paradigms often are synonymous, e.g., in the case of client-server systems (C-S). In addition, a complex service system could be implementing several control paradigms between its components: many web companies in fact are offering services implementing a highly distributed control paradigm between internal components and parts, and a simple client-server one for user interactions.

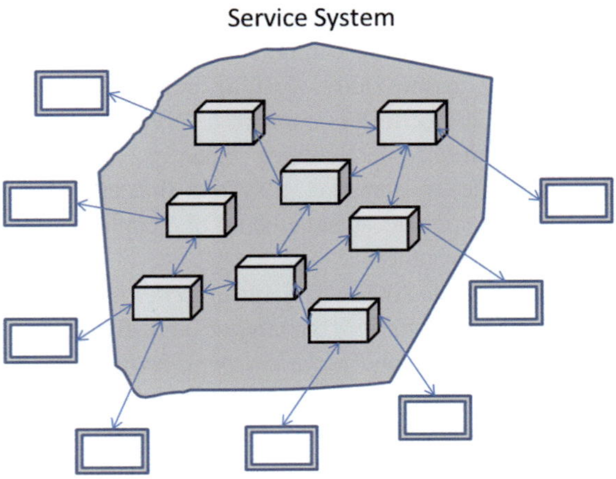

Fig. 2.3 External users' interactions with a service system

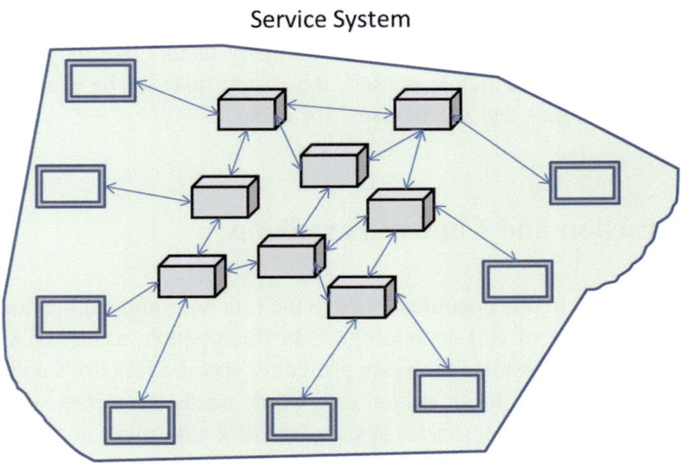

Fig. 2.4 Users as part of a service system

With reference to Figs. 2.3 and 2.4, the difference is relevant: in the first case, the relationship is clearly of a dependence, users can access the system functionalities that are owned, provided, and managed from a specific actor (the provider agent). In the second case, the ownership of the service system is blurred, since the users are providing functions and resources to the service system. It is difficult to see the users as external, without them the system would not even be able to exist. This is a great difference that points to the client-server (C-S) and the peer-to-peer (-P2P) interaction paradigms as represented in Fig. 2.5.

Graphically the difference is evident: in the client-server, the central role is played by the server (that usually is seen as the service system) that is the

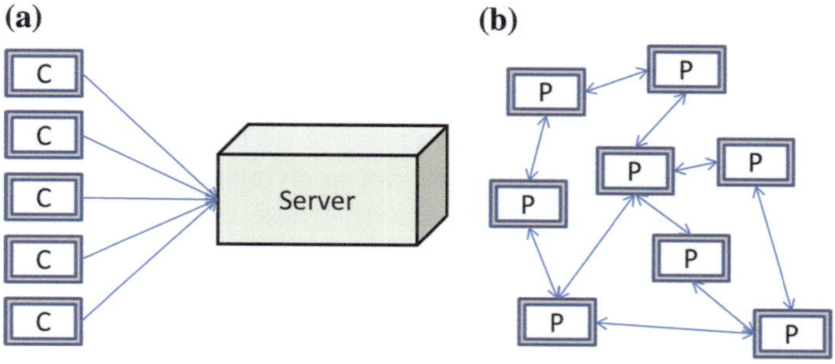

Fig. 2.5 Client–server and peer-to-peer interaction paradigms. **a** Client–server interaction paradigm. **b** Peer-to-peer interaction paradigm

centralized entity providing valuable functions to the clients that in these relationships have a lesser role. They can request functions and expect to receive a response. In the P2P paradigm, each single entity or node is equal to the others; there are not usually privileged relationships and each peer can request services from any other peer. Peers are designed and programmed for cooperation and the service system is totally decentralized. Also the communication mechanisms of P2P and C-S systems are different. In the P2P case, the communication between two communicating peers is direct in the sense that each peer knows the address of the other (even if the communication between them could go through other nodes, i.e., multi-hop manner), in the C-S case the communication is brokered, i.e., the server is in between the clients that do not have a direct reference of each other and are forced to pass through the server (Taylor and Harrison 2009). There is an extensive literature related to P2P and C-S, for example (Pavlopoulos and Cooper 2007; Orfali et al. 2007; Taylor 2005; Maly et al. 2003; Leopold 2001). Also the increasing wireless capabilities have raised the question of interaction and control paradigms (Datla et al. 2012).

The C-S paradigm is based on a very simple interaction between the clients and the server; a client sends a request (essentially a structured message) to a server and expects (nonblocking) a response from the server. Figure 2.6 depicts the simple interaction.

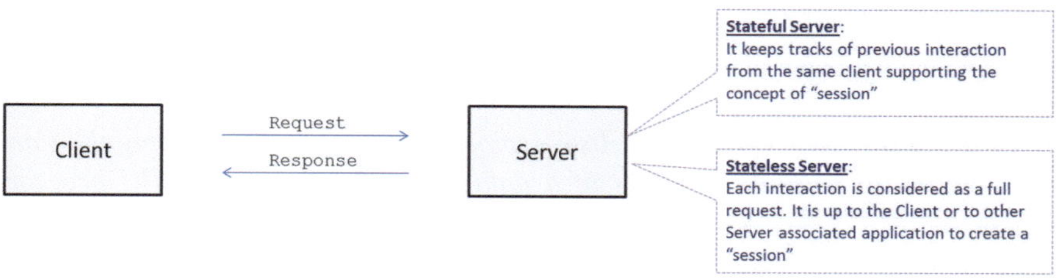

Fig. 2.6 The client–server interaction paradigm

The server is assumed to send a `response` to the client, however, the client cannot assume that it will be necessarily received. However, the `request` and `response` primitives are related and the client and server systems have to take care of this relationship. This means that a `response` cannot be received without a prior `request`. At the same time, a client not receiving the `response` could start polling the server in order to get an answer. These examples indicate that the interaction model is not balanced and there are several limitations. The server could be stateful or stateless, the difference is whether the system keeps track of the previous interactions with clients and has a finite state machine associated to the interactions going on. A stateful server is more complicated to manage especially if many clients are requesting in parallel the functions of the server. There is the REST architectural proposition (Fielding and Taylor 2002) related to web services and architecture that is promoting with a lot of success the idea of having stateless server as a principle of the architectural design.

Some interaction paradigms in proper sense are [in italics excerpts from (Foster 1995; Cachin et al. 2011)]:

- **Tasks and channels**. It is a computational model designed by Foster (1995) based on distributed tasks with logical channels connecting them. A task consists of a program, local memory, and a collection of input/output (I/O) ports. Tasks interact by sending messages through channels. A task can send local data values to other tasks via output ports. A task can receive data values from other tasks via input ports. Ports are connected by means of channels. The local memory contains the program's instructions and its private data.
- **Message passing**. *Message-passing programs create multiple tasks, with each task encapsulating local data. Each task is identified by a unique name, and tasks interact by sending and receiving messages to and from named tasks. The message-passing model does not preclude the dynamic creation of tasks, the execution of multiple tasks per-processor, or the execution of different programs by different tasks. However in practice, most message-passing systems create a fixed number of identical tasks at program startup and do not allow the tasks to be created or destroyed during program execution. These systems are said to implement a single program multiple Data (SPMD) programming model because each task executes the same program but operates on different data.* The difference with tasks and channels model is that a task needs only a queue to store messages and a queue can be shared among all the tasks that need to communicate with a specific task, while in the tasks and channels model, a channel (a queue) is created for supporting the communication between two tasks.
- **Data Parallelism**. *Data parallelism exploits concurrency by applying the same operation to multiple elements of a data structure. A data-parallel program consists of a sequence of such operations. As each operation on each data element can be thought of as an independent task, the natural granularity of a data-parallel computation is small, and the concept of "locality" does not arise naturally. Hence, data-parallel compilers often require the programmer to*

provide information about how data are to be distributed over processors, in other words, how data are to be partitioned into tasks. The compiler can then translate the data-parallel program into an SPMD formulation, thereby generating communication code automatically.

- **Shared Memory**. *In the shared memory programming model, tasks share a common address space, which they read and write asynchronously. Various mechanisms such as locks and semaphores may be used to control access to the shared memory. An advantage of this model from the programmer's point of view is that the notion of data "ownership" is lacking, and hence there is no need to specify explicitly the communication of data from producers to consumers. This model can simplify program development. However, understanding and managing locality becomes more difficult, an important consideration (as noted earlier) on most shared memory architectures. It can also be more difficult to write deterministic programs.*

The message-passing model is represented in Fig. 2.7. It is based on a very simple organization in which a sender forwards messages by means of a queue to a recipient.

Obviously the different mechanisms have advantages and drawback. For a deeper analysis (Kubiatowicz 1998; Foster 1995) provide more details and discussions. A short analysis of pros and cons is provided in the following:

- Message Passing: Message passing (MP) paradigm can be interpreted as an "interrupt with data," i.e., events are queued for processing together with associated data (they could also be large bulks of data). Receiving tasks can then operate on received data when the message is taken from the queue. There is the need of an explicit treatment of communication (e.g., "send message to task *n*") and management of data (data are copied from data structures in the source of the message, transmitted to a queue and then copied to a data structure at the sink destination). Assembling and disassembling of messages can be cumbersome.
- Shared Memory: one of the major advantages is that the communication details between programs and the "shared memory" are hidden to the programmers, freeing them from the problem of dealing with communication (in the shared memory) and the data management (replication, caching, data coding and encoding, optimization of distribution of data). The programmers can actually focus on the parallelism and the issues to solve. However, the programs have to

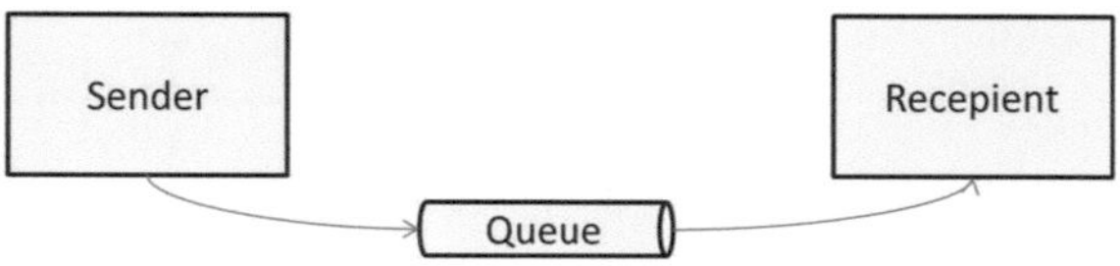

Fig. 2.7 Message-passing model

adopt a sort of polling in order to synchronize with new data and new events (and this can be time and resource consuming), in addition data management is optimized having in mind a model optimization. Specific applications needing other kinds of optimization or a hybrid composition between data parallelisms and other approaches will find it difficult to implement specific application oriented policies in this context. This could be an example of abstraction that eases programming, but sometimes it is detrimental for applications that need to do "different" things.

- Tasks and Channels: this paradigm can be seen as a variation of the message passing. According to Foster (1995), the "task/channel" paradigm enforces the programmer to better conceptualize the communication of a parallel and distributed program. This could provide advantages in designing a large distributed system, but it also introduces more overhead in the need to control individual channels and in creating a relationship between channels and related tasks.
- Data Parallelism: while this paradigm can lead to significant advantages when the same instructions can be applied to different data, its general application is not easy. In fact not all the application domains offer problems that can be effectively solved with this paradigm. Actually the major drawback of this paradigm is its applicability to more restricted set of applications compared to the previous ones.

These systems are to be analyzed and used appropriately according to the specific needs of the applications. In the rest of this section, an analysis of their benefits, merits, and applicability is carried out. In large MP systems, senders and recipients share a common infrastructure made out of queues and these can be organized as a sort of message broker in charge of receiving and dispatching (in an asynchronous manner) messages between sources and sinks (Fig. 2.8).

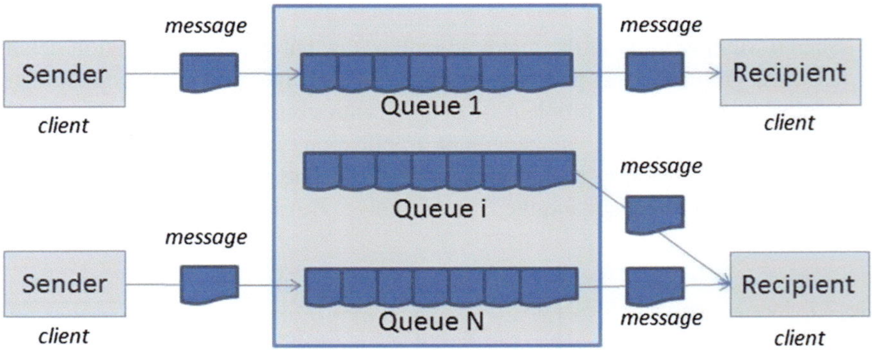

Fig. 2.8 A message-passing (MP) system

There are at least three important features.

- The model can be synchronous or asynchronous, i.e., in the first case, the sender waits for the receiver to deal with the message. In the second case, once a message has been sent, the sender can proceed with its computation. The choice of how to control the timing of operations (in a synchronous or asynchronous manner) depends on the applications requirements. The introduction of queues to store messages for later processing is a case in which asynchronous mechanisms are implemented.
- Routing: the system (especially by means of the Broker Functions) can route the messages even if the sender has not provided a full path to the destination.
- Conversion: the Broker functions can also make compatible different messages format in such a way to support the interoperability between different messaging systems.

Communication between the Sender and Receiver is not "brokered", because the event message manager enables an asynchronous but full communication between the two parties.

The message passing, MP, and the C-S paradigms have commonalities: i.e., a sender and a receiver can be easily mapped on a client and a server. However in a MP system, the role of sender and receiver are less constraining than in a C-S system. In fact, the MP model is more granular (in a sense that the sender could be just sending messages—events—to a receiver without expecting any "response" or any service from the server). In the MP paradigm, messages are not necessarily correlated between a client and a server, and actually each entity can be a sender (and not necessarily a receiver). Under this perspective, the MP paradigm can be used in order to support a C-S model: `request` and `response` primitives can be implemented as two separate messages exchanged between a sender and a recipient. On the other side, using a C-S system for implementing a MP system introduces the logical concept that the receiving party will offer a service to the sender or at least will respond to the received message. This means that the expressiveness of the MP is greater than the C-S paradigm even at the cost of further complications[2] (Fig. 2.9).

The P2P paradigm is very close to the Message Passing as well as the Task and Channel models. Each peer can indeed be seen as an entity capable of being a sender and a receiver where each of them uses channels for sending and receiving data (messages, events, and flows). Actually some P2P protocols, e.g., Gnutella (Ripeanu et al. 2002), seem to be designed having the MP paradigm in mind while others are taking the Task/Channel model as the founding concept: e.g., the i2p tunneling mechanisms (Aked 2011).

---

[2] Actually a double C-S system could provide the functionalities of a MP paradigm. However the difference between the server side and the client one is such that many clients are not capable to host both the client and the server functions.

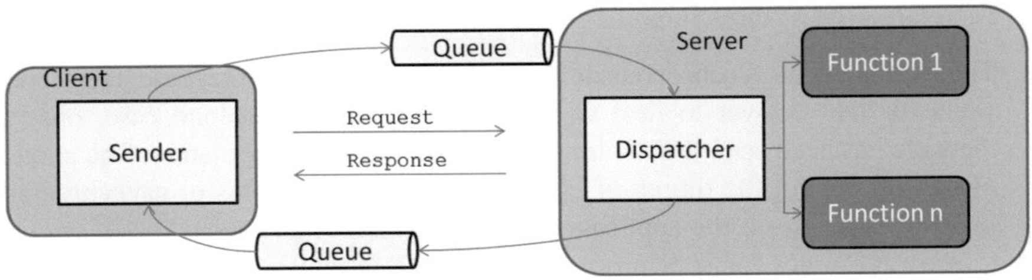

Fig. 2.9 A C-S relationship implemented by means of a MP system

Other interaction mechanisms are emerging especially for data intensive processing, for instance (Grelck et al. 2010; Margara and Cugola 2011) analyses the relations between stream processing and the Complex Event Processing.

2.4 Network Architectures and Distributed Platforms

Even if the interaction paradigms are independent from the network capabilities, in reality they are strongly influenced by the underlying communication infrastructure in several ways. As stated by Deutsch (1995) and further elaborated by Rotem-Gal-Oz (2006), (Thampi 2009), there are network fallacies that should not be ignored in designing the networked applications. Robustness and resilience of networks cannot be taken for granted and hence service design should take care of utilizing interaction, and control paradigms capable of coping with these fallacies as well as the right partition of functions within different administrative domains (including users' domain).

Another important notion related to distributed platforms is their programmability. Programmability is exerted by means of protocol interactions or more and more frequently in the Web and IT world by means of application programming interfaces (APIs). An API (Magic 2012; API 2013), is a source code-based specification intended to be used as an interface by software components to communicate with each other. An API may include specifications for routines, data structures, object classes, and variables. A difference between an API and a protocol is that the protocol defines a standard way to exchange requests and responses between functional entities (to be seen as black boxes) while APIs expose a part of the software infrastructure (and organization within the functional entity). An API can be

- language-dependent, meaning it is only available by using the syntax and elements of a particular language, which makes the API more convenient to use.
- language-independent, written so that it can be called from several programming languages. This is a desirable feature for a service-oriented API that is not bound to a specific process or system and may be provided as remote procedure calls or web services.

2.4.1 *Distributed Platforms as a Means to Cope with Network Fallacies*

As seen for the message passing MP paradigm, there is a need to design how messages (or events) are passed between the involved entities. In addition, some extra entities besides the communication ones are usually involved in the interactions. For instance, queues and message brokers are networked elements supporting the interaction paradigm. The client-server paradigm strongly relies on the capabilities of supporting communication protocols. There are two major flavors of the client-server paradigm: the IT one and the Web one.

2.4.1.1 The Client–Server Paradigm as Supported by IT Technologies

The first one is based on remote procedure call (RPC) mechanisms as depicted in Fig. 2.10.

A few observations are useful: if the client and the server functions are synchronous, the underlying mechanisms are based on events/messages. In fact, the network drivers will fire events and messages as soon as they arrive; the underlying network is based on IP communications with two options of connection: connection-oriented (e.g., Transmission Control Protocol, TCP) or connectionless (e.g., User Datagram protocol, UDP). This means that requests can pop up asynchronously and the network driver is essentially a sort of internal queue manager; the upper layer functions are usually provided by means of application programming interfaces, API.

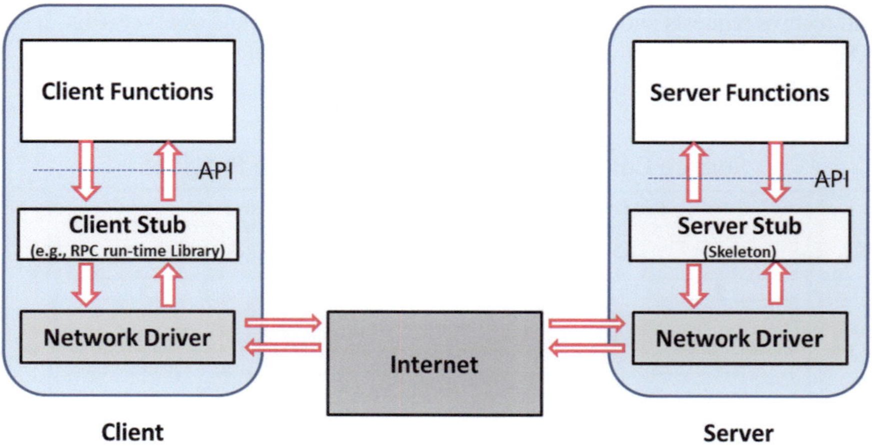

Fig. 2.10 The client–server paradigm and remote procedure call mechanisms

This approach paved the way toward middleware platforms. The initial RPC systems were in fact proprietary and confined to well-defined operating systems with machines essentially interconnected by enterprise networks. From this stage, the introduction of middleware platforms, i.e., software functions in between applications and the operating system capable of creating a sort of (distributed) unified platform supporting these functions as stated in Krakowiak (2003):

- *Hiding distribution, i.e., the fact that an application is usually made up of many interconnected parts running in distributed locations*;
- *Hiding the heterogeneity of the various hardware components, operating systems, and communication protocols*;
- *Providing uniform, standard, high-level interfaces to the application developers and integrators, so that applications can be easily composed, reused, ported, and made to interoperate*;
- *Supplying a set of common services to perform various general-purpose functions, in order to avoid duplicating efforts and to facilitate collaboration between applications.*

This approach was the basis for DCOM (Krieger and Adler 1998) and CORBA (Object Management Group 2006; Chung et al. 1998; He and Xu 2012). More recently, this approach has led to the definition of the OSGi platforms (Kwon et al. 2010, 2011) as represented in Fig. 2.11.

2.4.1.2 The Client–Server Paradigm as Supported by Web Technologies

The other trend within client-server paradigm is based on Web technologies. In this case, the basis is the Hypertext Transfer Protocol (HTTP) used as a means to send and receive requests and responses between a remote client and a server (Fig. 2.12). XML_RPC (Laurent et al. 2001) is a very simple and effective protocol for C-S

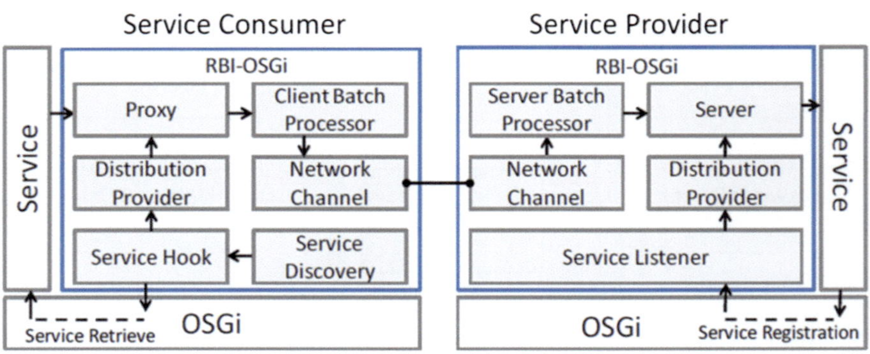

Fig. 2.11 RBI/OSGi architecture (Kwon et al. 2011)

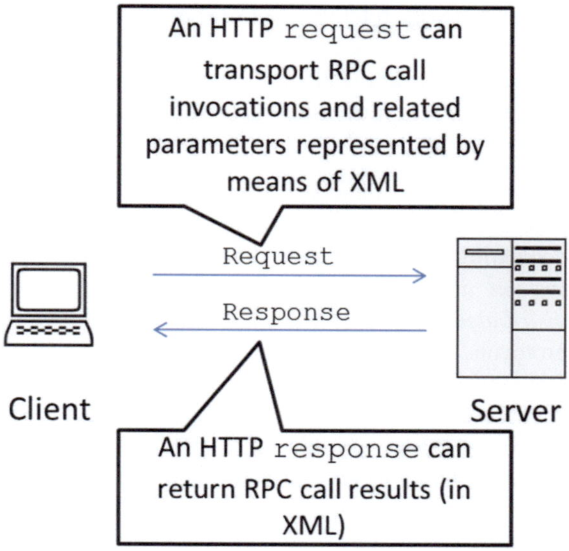

Fig. 2.12 RPC interactions based on HTTP

interactions based on HTTP for transport and Extensible Markup Language (XML) for information representation. Remote procedure invocation and responses are coded with XML and the payload is forwarded by means of HTTP.

The simple idea that an RPC mechanism can be built using XML and HTTP (Richards 2006) has started a new view of creating web services, i.e., the possibility to dynamically invoke services in the web directly from a client remote application. This approach has led to the definition of new protocols such as Simple Object Access Protocol (SOAP) as well as to new approaches in the software architectures (e.g., REST). For a while these two approaches diverged: a lot of effort was put on the specification of Web Services Architecture and Service-Oriented Architectures. However, the increasing importance of the Representational State Transfer (REST) approach has suggested the need to integrate the two approaches under the web services definition (see Sect. 2.5).

2.4.2 Control Paradigms and Network Architectures

The combination of the communication issues, i.e., how communications capabilities are supported, offered, and used by the systems that implement services, and the interaction paradigms themselves define a set of control paradigms

- the client-server paradigm that exploits protocols, architectures, and capabilities of the Web and IT industry in order to support services by means of networked resources;

- the Network Intelligence (NI) paradigm, that combines event-based processing with the network capabilities and resources;
- the peer-to-peer (P2P) approach that combines the message passing mechanisms (or the event processing one) with distributed capabilities offered and controlled by the different peers over an overlay network.

As stated in Santoro (2006), the C-S paradigm is not properly a distributed processing paradigm: "*Incredibly, the terms "distributed systems" and "distributed computing" have been for years hijacked and (ab)used to describe very limited systems and low-level solutions (e.g., Client-Server) that have little to do with distributed computing*". The C-S paradigm can be seen as a by-product of a set of protocols, functionalities and mechanisms that allow the distributed communication, interaction and coordinated processing. The C-S paradigm does not really contribute to the advancement of distributed processing, it just relies on those techniques. However, it is considered here for its profound impact and importance in the evolution of service platforms.

Services are built around interaction and control mechanisms. Users can access and take advantage of service functionalities by interacting with a system. For instance the C-S one is very simple but it requires the "client" to activate the service (pull) and wait for a response. Generally, it is not envisaged that a "server" will autonomously initiate an interaction toward the user (a push). This is obviously a limitation, services are designed in such a way to adhere to this interaction paradigm, and not all the services can be appropriately provided in a—C-S fashion. For instance, a simple service like "inform the client when a condition on certain data is met" (event notification) can be tricky to implement in a C-S fashion, but it could be simple to realize with another paradigm (e.g., a P2P implementation based on a message passing mechanism). The issue with the C-S paradigm is a well-known one [see for instance (Adler 1996): "Important examples include task allocation and event notification in collaborative workgroup systems, and task sequencing and routing in workflow applications"] and it has an impact especially when users/tasks need to coordinate several threads of processing and information distribution. Sometimes, in order to circumvent this issue, a continuous polling of the client on the server can be implemented, but this is not an optimized solution. Solutions like Websockets (Hickson 2011) can be seen as a way to solve this problem in the context of HTTP client server communication. Here a socket is created and the server can forward on the socket asynchronous information. In this way, the "difference" between the client and the server is reduced and this architecture can be seen as a move toward P2P solutions also for the Web. The other paradigm, the Network Intelligence one, is even more limited. It is based on the network capability to determine that the user is requesting service functionalities. The network "control" then triggers (i.e., it requests for assistance to a set of service layer functionalities) to the specific service. In this case, the service is confined to the network context in which the user is operating which is also confined by the protocols and mechanisms of the network. The P2P paradigm seems to be the most adaptive, it supports a granular message-based (or event processing) approach that allows to design the service functions very effectively at a cost of representing a logical network of relationships between peers.

The paradigms have different “expressive power” (Minerva 2008a, b, c; Minerva et al. 2011), i.e., different capabilities to represent and support the control needs of a service. Actions and interactions within the parts of a service system can be represented and supported in different fashions and with different mechanisms, expressive power means here how easily the control paradigm can represent and support the needed communications and control of the components. Introducing functions to circumvent problems that could be easily solved with other paradigms could be a sign of low expressive power. Examples are the polling mechanisms for C-S systems compared to message passing systems or event-driven programming, or the over simplification of ParlayX APIs that do not allow the support of anonymity in services. An exemplification of the lack of expressive power in C-S paradigm is represented by WebSocket definition (Fette and Melnikov 2011): a new mechanism that extends the paradigm capabilities had to be introduced in order to allow the server side to autonomously notify the occurrence of events of partial data to the client side. Lack of expressive power can also yield to phenomena called Abstraction Inversion: users of a function need functionalities implemented within that are not exposed at its interface, i.e., the need to recreate functions that are available at lower levels. Abstraction Inversion, too much abstraction, and wrong control paradigm can lead to anti-patterns in intelligent systems (Laplante et al. 2007), with consequences in the ability of platforms to support services, increase in cost of software development, longer processes to delivery phase. Generally put, if a control paradigm requires too much adaptation and development of specific functions for representing and support the control flows which means that the paradigm is not well suited for such service.

The issue of expressive power is a fundamental one: the possibility to implement services with the best model for supporting the interactions between its parts should have a large importance when platforms and systems are designed, implemented, or simply bought. Instead it is often neglected and service offerings are created on the available platform totally disregarding how the final service will be provided. This has led (especially in the telecommunication world) to aberration in which highly interactive and cooperative services have been implemented with a very constraint paradigm (Minerva 2008b, 1) such as Network Intelligence [see for instance the attempt to implement Video On Demand (VoD), over IP Multimedia Subsystem (IMS) (Riede et al. 2008)]. The Interaction and Control paradigms are at the very heart of the service design and implementation, and this issue is one of the crucial considerations for determining how to create effective service platforms.

The considered control paradigms are depicted in Fig. 2.13 and are briefly discussed in this section.

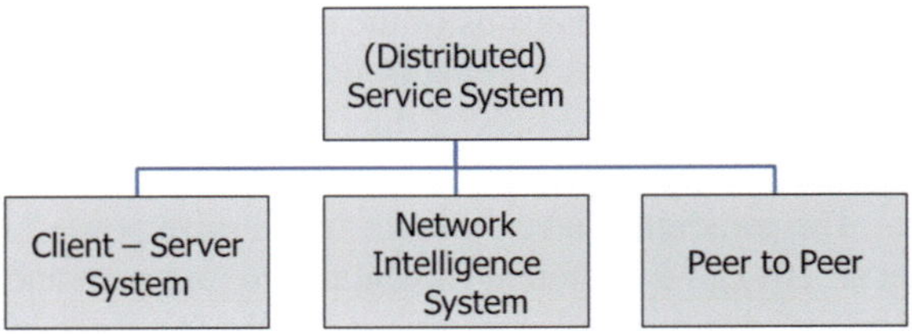

Fig. 2.13 Control paradigms in the context of networked interactions

2.4.2.1 The C-S Control Paradigm

The C-S control paradigm is the simplest one; it is based on the concept that the network is essentially transparent to the functional interactions between the clients and the server. It clearly and cleverly exploits the end-to-end principle (Saltzer et al. 1984) of the Internet in order to opportunistically use the network capabilities. This principle states that "*mechanisms should not be enforced in the network if they can be deployed at end nodes, and that the core of the network should provide general services, not those tailored to specific applications*". So the C-S paradigm is a means to move the intelligence at the edges of the network and confining the network to provide generic and reusable services related to transport. This has given rise to the concept of stupid network as proposed by (Isenberg 1998) Fig. 2.14 represents the concept of the stupid network.

As stated in Minerva et al. (2011), the C-S control paradigm has several merits

- Simple control pattern
- Implemented in stateless or stateful manner
- Decoupling of application functions from the network intricacies (service **deperimeterization**).

Its major drawback is that not always the network has an ideal behavior and there is often the need to orchestrate resources and capabilities in order to provide compelling services.

2.4.2.2 The Network Intelligence Control Paradigm

It is the typical control paradigm used by the telecoms industry to provide services. It leverages the capabilities of the network to provide services. Figure 2.15

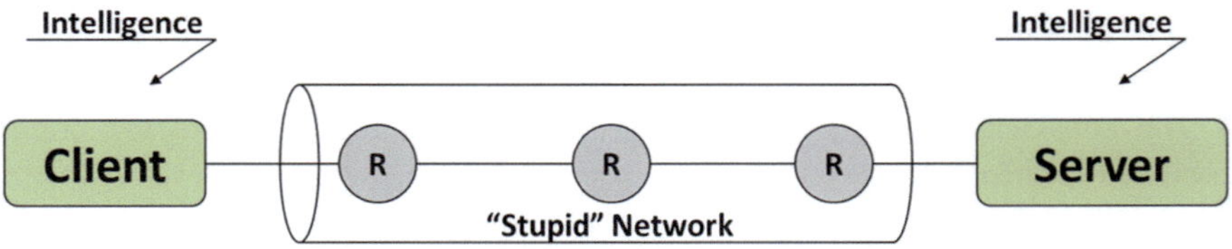

Fig. 2.14 The stupid network and the aggregation of intelligence at the edges

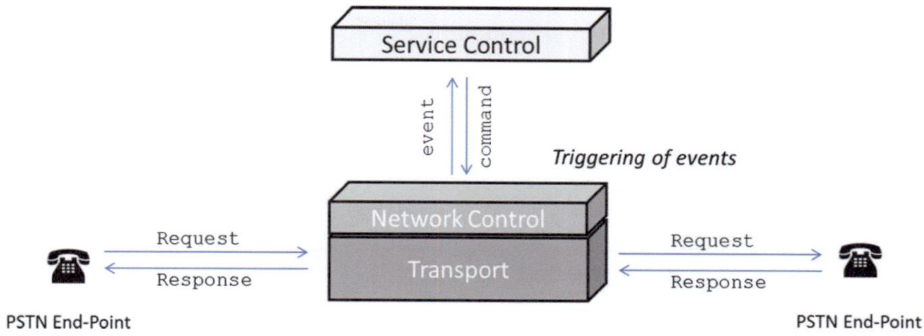

Fig. 2.15 Triggering in the NI paradigm

illustrates the basic mechanisms behind. In this case, the clients can interact with the control functions of the "network" and only when it is not possible to serve the request, i.e., when the service is "triggered". There are two interaction models coupled in this control paradigm: a sort of C-S between the users of the "network service" and an event-driven model (Dabek et al. 2002) for the interaction between the control functions and the (added value) service functions.

The composition of interaction model is a bit complicated in this control paradigm. The caller (or the initial requestor) could be considered in a stateful C-S relation with the network. The network control triggers events toward a service control entity when it has no instructions on how to proceed with the service request. This relation is not Client-Server, because more (correlated) events can be sent by one party to the other. It is more an event-driven approach. Also the called or the destination of the communication is in an event-driven relationship with the "service system". One evidence is that the network control is in charge of the orchestration of services. There is no way for the end points to request directly services and functions to the service control. This layer is mediated by the "network". This approach was meaningful when the end points were stupid, but nowadays, terminals have plenty of intelligent capabilities and they could be capable of benefitting from a direct interaction with the service control and even from the direct communication with network resources. This dependence of the services from the network is a major drawback. In addition, the complexity of this model is even exacerbated by the fact that all the interactions are stateful and organized around very complex Finite State Machines (FSM) (Minerva 2004). The FSMs are associated to the concept of call and more recently session control model. Typically a service progresses along the states and the possibilities offered by a very complex and controlled model does not offer too much flexibility to the programmers (Minerva 2008b, 1), in addition the call and sessions are strongly coupled with network resource allocation. For instance in Fig. 2.16, yellow resources are allocated to a session.

Resources are made available only within a session in order to access another resource, the session control has to negotiate for it, while the end points have only the possibility to interact with the session control to request a mediated control of

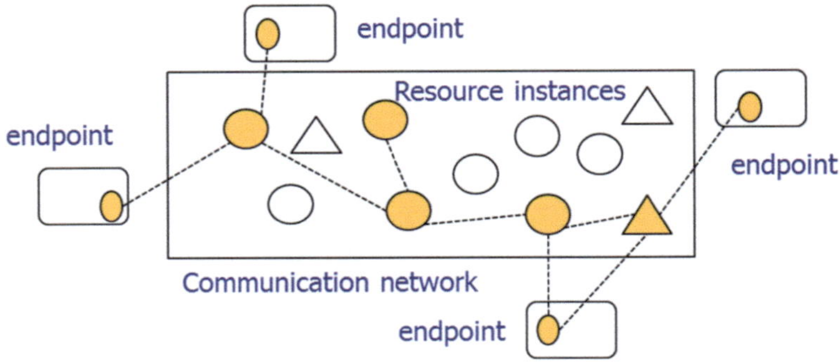

Fig. 2.16 The session concept and the allocation of resources

new resources. The possibility for endpoints (with increasing level of intelligence) of direct interaction with a specific resource is totally neglected by the Network Intelligence control paradigm. This leave outside of the service control those elements that have the major requirements to do so.

This approach has been successful for the Intelligent Network (Faynberg et al. 1996), and it has been reiterated for many of the new control architectures for telecoms such as IP Multimedia Subsystem, IMS, (Camarillo and Garcia-Martin 2007), and Next Generation Networks in general (Copeland 2009). The IMS in particular has been considered as the new control platform to be used in order to offer the integration between telecoms and Internet services. This assumption is based on the following features offered by the IMS:

- Multi-access Networks, i.e., the capability to support several heterogeneous access networks
- Layering and decoupling of functions at the transport, control and service level
- Reuse of the successful general packet radio service (GPRS) functional architecture.

IMS is then seen as a reference architecture for Next Generation Networks. This approach has some pros and cons, such as

- Regulated and Standardized
- Support of interworking
- Constrained and limited
  - New add-ons are difficult to introduce in the software
  - Stateful model
  - Session control is just an evolution of the call control
- Complexity increases with number of new functional entities and states (for instance the presence of different types of Call State Control Functions (CSCF), gives rise to the need of several protocol interaction in order to sort out the Functions orchestrating the service).

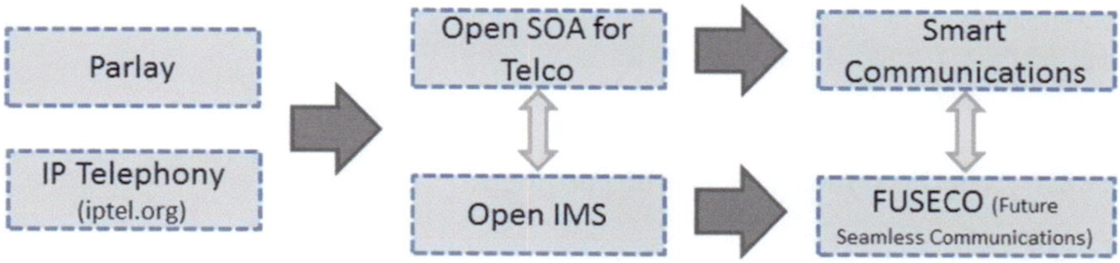

Fig. 2.17 The NGNI playground evolution

This is a very traditional approach that disregards the issues in interaction and control paradigms, the evolution of terminals and edge functionalities, the current development of the major projects in the Future Internet such as GENI (Peterson and Wroclawski 2007), AKARI (Aoyama 2009) and others (Pan et al. 2011). These new architectures consider different interaction and control paradigms (highly distribution of functions, P2P, mesh networking) that greatly differ from traditional ones. For a good overview of Network Intelligence, the interested reader could refer to the ICIN conference www.icin.co.uk that collects many contributions in this field. For a traditional approach and its mainstream developments in the exploitation of Network Intelligence, good sources are (NGNI Group 2005, 2006, 2009, 2012). Figure 2.17 represents a (desired) mainstream evolution of technologies and system from a telecoms perspective.

Summarizing, the pros and cons of the NI approach are

- Mix of different mechanisms (C-S and event—commands)
- Generally implemented in a stateful manner
- Strong dependence of application functions from networks ones (e.g., the triggering mechanism is the way in which services can be activated, creating a strong dependence of services on the status of the network).

Some other major concerns in the NI approach are: the **perimeterization** of services (i.e., services are tightly coupled to the network functionalities and its geographical extension) and the assumption that Quality of Service (QoS) is a major requirement for service. The definition of the IMS architecture can be considered as an example of the importance given by the telecommunication industry to the introduction of dynamic allocation mechanisms for better controlling the bandwidth allocation, and related QoS parameters. There is a continuous strive to find out models to introduce in the network these kinds of mechanisms (Song et al. 2007). They have been discussed and criticize in Minerva and Bell (2010) and Minerva et al. (2011) showing how operators cannot compete at a global level because their services cannot be provided outside of the deployed and owned network and arguing that QoS and more bandwidth are not always practical solutions for solving service issues (in fact, Google solved the problem of long map downloading time by introducing AJAX solutions based on asynchronous download of data[3]).

2.4.2.3 The Peer-to-Peer Control Paradigm

Apart from the different topologies and characterization of structured, unstructured, and hybrid peer-to-peer networks (Lua et al. 2005; Merwe et al. 2007), two concepts are central to P2P networks: equality of peers, i.e., each node is potentially a client and a server of the network; overlaying and virtualization, i.e., the creation of a logical virtual network on top of physical resources. Another important aspect is the dynamicity of the network: nodes leave and join dynamically and the average uptime of individual nodes is relatively low. The topology of an overlay network may change all the time. Once a route is established, there is no guarantee of the length of time that it will be valid. Routing in these networks is therefore very problematic. In other terms, the P2P control paradigm is complicated by the need to manage and keep track of a highly dynamic infrastructure that changes over time. On the other side, the interaction between peers could be implemented adopting the most appropriate (for the specific service) interaction mechanism. So a P2P infrastructure could be able to encompass several interaction paradigms offering to services the most suitable mechanisms.

P2P network Systems are characterized by

- Complexity of the entire system and simplicity of each node
  - Nodes are both clients and servers
  - P2P systems are real distributed systems, they need to govern an overlay network (resource naming, addressing, location)
- Scalability and Robustness
  - P2P systems can scale to millions of nodes
  - Problems in detecting the source of issues in case of malfunctions
- Different way of programming
  - P2P systems can be implemented as event-driven system but also the communication between two terminal nodes can be client-server
- Social flavor of the solutions (Koskela et al. 2013). The willingness to share and the thresholds at which an altruistic behavior is triggered have been studied

[3]Actually this is another evidence that the expressive power of the C-S paradigm is not sufficient to provide services that do require responsive control on the flow of data. Introducing asynchronous (and invisible to the users) calls between the client and the server is pointing to the need for more capable control mechanisms. Actually those mechanisms are also provided in WebSocket that definitely change the C-S paradigm (in fact the server can use now the socket to notify events and messages to the client).

(Ohtsuki et al. 2006). They are particularly meaningful to understand when a P2P network will become valuable to a single user.

Summarizing, the P2P approach seems to be very flexible and promising; some pros and cons are

- Able to support many control patterns
- Generally implemented in a stateful manner (the overlay control)
- Intertwining of application functions with network ones.

2.5 Service Architectures

The interaction and control mechanisms have to be framed and supported by software architectures. In this context, a software architecture consists of a set of concepts, principles, rules, and guidelines for constructing, deploying, operating, and withdrawing services. It also describes the environment in which such services operate. The service architecture identifies components to build services, describe the way they are combined, and the way they interact. The combination of organization principles of the architecture, the underlying technology platform and the enabled interaction and control paradigms determine the expressive power of the service architectures and its capabilities in supporting several classes of services. The expressive power is the capability of services and application to represent the real world interactions and to organize the available functions in such a way to accommodate the applications and users needs.

There are some major attempts to define software architectures (Proper 2010; Lankhorst 2013) and their expressive power with respect to the problem domain representation and the involved stakeholders.

2.5.1 The Web and IT Views on Service Architectures

The evolution trajectory of Web and IT service architecture is defined by different factors such as

- Evolution of web servers toward application server architectures (Minch 2009);
- Evolution of Web Service (Booth et al. 2004) and Service-Oriented Architectures (Partridge and Bailey 2010) and the competition between SOAP and REST based solutions;
- The development of advanced solutions within data centers that have led to Cloud Computing (Gong et al. 2012) opposed to Grid Computing (Foster et al. 2008; Sadashiv and Kumar 2011);

- Virtualization of resources (Chowdhury and Boutaba 2010; Sahoo et al. 2010) and the Software as a Service models (Prodan and Ostermann 2009).

An attempt to standardize the mechanisms behind the simple XML_RPC approach has been done with [Simple Object Access Protocol (SOAP) (Box et al. 2000)]. SOAP provides a simple and lightweight mechanism for exchanging structured and typed information between peers in a decentralized, distributed environment using XML. SOAP consists of three parts

- The SOAP envelope defines an overall framework for expressing what is in a message, who should deal with it, and whether it is optional or mandatory.
- The SOAP encoding rules define a serialization mechanism that can be used to exchange instances of application-defined data types.
- The SOAP RPC representation defines a convention that can be used to represent remote procedure calls and responses.

The definition of SOAP was instrumental to the growth of the Web Services Architecture. Web services provide a systematic and extensible framework for application-to-application interaction, built on top of the existing Web protocols and based on open XML standards (Curbera et al. 2002). The Web services framework is divided into three areas: communication protocols, service descriptions, and service discovery. They correspond to three protocol specifications that are the cornerstones of the Web Services Architecture (Curbera et al. 2005):

- the mentioned simple object access protocol which supports communication between Web services;
- the Web Services Description Language (WSDL) (Chinnici et al. 2007), which specifies a formal, computer-readable description of Web services;
- the Universal Description, Discovery, and Integration (UDDI) (Clement et al. 2004) directory, which is a registry of Web services descriptions.

These protocols form the skeleton of the well-known Web Service Architecture represented in Fig. 2.18. The Service Provider uses the UDDI functions in order to register and advertise its services on the Universal Registry. A service requestor can access the UDDI functionality and search for the most suitable service for its needs. It gets a WDSL description that can be used in order to correctly access the service. At this point, the service requestor and the service provider can be linked and can use the SOAP Protocol to access and provide the service functions.

The specification has been extensive and the original simplicity of the solution has led to a complex set of documents and to overall complex implementations. Essentially the UDDI service, originally intended to be used in a dynamic way, is used off-line in order to keep track of the software components made available. The SOAP protocol itself has been disregarded in favor of the simplest solutions based on the REST approach (Fielding and Taylor 2002). REST is a set of architectural principles for designing and implementing distributed systems. It proposes

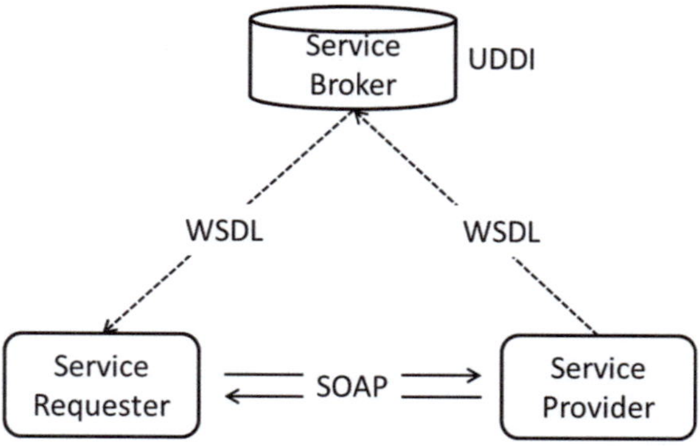

Fig. 2.18 The web services architecture

the use of web-based interfaces that are described in XML and handled through HTTP. The REST principles are:

- A stateless client–server protocol
- A set of well-defined operations (e.g., GET, POST, PUT, and DELETE) to perform functions on resources
- A universal syntax for resource identification (resources are identified by URIs)
- Use of XML or HTML for describing pieces of information and their links.

REST is at the core of Web Application Development as shown in Fig. 2.19. This figure shows that whenever a simple solution is offered and it is based on well-known mechanisms then the programmers will adopt it. This is the case of REST: it is based on the established principles of web programming (each function is an addressable resource and the mechanisms to interact are those supported by the HTTP) and it is simpler than other protocol and related solutions (like SOAP). Web programmers use it much more than other solutions.

The Web Service Architecture has mingled with the definition of service-oriented architectures (SOA) (Sward and Boleng 2012; SOA-Wikipedia

Fig. 2.19 Protocol usage by APIs. *Source* http://programmableweb.com

Protocol Usage by APIs	Percentage
REST	63 %
SOAP	22%
JavaScript	7%
Atom	3%
XML-RPC	3%
Others	2%

2013). Service-oriented architecture is a software architectural concept that defines how to use services to support the requirements of software users. In a SOA environment, nodes on a network make resources available to other participants in the network as independent services that the participants access in a standardized way.

Most definitions of SOA identify the use of Web services (i.e., using SOAP or REST) in their implementation. However, SOA can be implemented using any service-based technology.

Unlike traditional object-oriented architectures, SOAs comprise loosely coupled (joined), highly interoperable application services. Because these services interoperate over different development technologies (such as Java and .NET), the software components become very reusable, due to the virtue of the interface definition being defined in a standards compliant manner (WSDL) which encapsulates/hides the vendor/language specific implementation from the calling client/service. The relations between SOA and Web Services were first of competition and then of integration, nowadays SOAs can be implemented as Web Service Architectures.

The availability within data center of storage and processing capabilities has led to the development of Cloud Computing, i.e., "*a large-scale distributed computing paradigm that is driven by economies of scale, in which a pool of abstracted, virtualized, dynamically scalable, managed computing power, storage, platforms, and services are delivered on demand to external customers over the Internet*" (Foster et al. 2008). This concept seems to be very close to the grid computing. Very similar in scope, cloud computing and grid computing differ in at least two features that are important for this document

- The standardization and openness of the approach
- The control and optimization of the underlying resources with special emphasis on the network ones.

The grid, in fact, has pursued a standardization of its solution from its incipit and it has considered resources with holistic and broad view. Figure 2.20 represents the

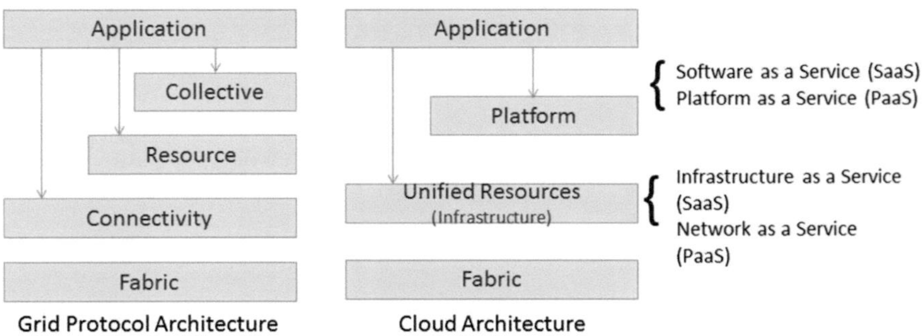

Fig. 2.20 Grid and cloud computing architectures

two different approaches: the grid aims at controlling resources and groups of them also deployed in different administrative domains, while the cloud (at least originally) was aiming to homogeneous and single administrative domains. In addition, the grid is aiming at granular control on the resources and it also fosters the control on connectivity as one major goal.

Only recently, cloud computing (and the Web Companies behind it) have started to work with the goal to achieve a better control of connectivity. For instance Google has unveiled their effort to build a large-scale controllable network (Vahdat 2012).

Contextually to the development of cloud computing, virtualization technologies have made such progresses to become a major technology trend in several application domains. Resource virtualization is the possibility to create a virtual machine that acts and behave like real resources on top of hardware and software resources. Software executed on these virtual machines is separated from the underlying hardware resources. The benefits reside on the possibility to decouple even further the software infrastructure from the underlying hardware and to segment on a cluster of machines different applications and resources limiting the occurrence that the fault of a (virtual) machine could impact on other machines and resources (Metzler 2011). Virtualization was instrumental to the possibility to introduce cloud computing (Lenk et al. 2009) and the Software as a Service (Turner et al. 2003) model to a large audience. Its extension has led to the new trend of virtualizing everything as a service [XaaS, (Schaffer 2009)] and to offer these features by means of a cloud infrastructure.

Summarizing, the Web and IT world is aiming at very large structures that can virtualize processing, storage resources, and can offer compelling services developed and organized in reusable building block. Resources are addressable with web mechanisms and they offer APIs that can be mashed up and combined in order to provide new services and applications. In spite of standardization, these solutions aim at proprietary environments in order to capitalize a technology investment and advantage.

2.5.2 *The Network Intelligence View on Service Architectures*

The evolution of service architecture in a Network Intelligence context is characterized by a two competing approaches and visions

- the traditional Telco one that can be seen as an evolutionary path from the ISDN, IN, and IMS architectures;
- an innovative one that stems from open platforms and in particular TINA and goes through Parlay and the service delivery platform in order to leverage Network APIs.

The two approaches have often been competing for driving the evolution of the Next Generation Networks, NGN, but nowadays they stick together because they have become rather traditional and generally obsolete. The windows of opportunity for opening up network interfaces, building a new programmable architecture and offering new services was around 2000–2002 (along the initial deliveries of Web 2.0 platforms). After that period, the window of opportunity closed as discussed in several papers that have identified the decline of the network intelligence (Minerva et al. 1998; Minerva and Moiso 2000; Licciardi et al. 2000, 2004; Manzalini et al. 2009).

The current attempts to repositioning the Operators in the Web 2.0 wave (Maes 2010) are probably useless and are consuming resources that should be spent for other efforts (see for instance Telco as a Platform Enabler and Disappearing Telco, to be discussed in Minerva and Crespi (2016)).

In the Telecommunication world, a lot of effort (in terms of joint projects, research activities and association of several enterprises) has been spent on the definition and demonstration of viable programmable platforms. A few contributions touching relevant issues posed by the book with particular emphasis on new telecom architectures are listed below:

- Architectural definitions for TINA (Berndt and Minerva 1995; Berndt et al. 1999; Yagi et al. 1995), Parlay (Walkden et al. 2002), Open Service Access (OSA) (Moerdijk and Klostermann 2003) and Service Delivery Platform, SDP (Moriana Group 2004; Baglietto et al. 2012);
- Definition of Open Interfaces like Parlay (Di Caprio and Moiso 2003), ParlayX (Parlay Group 2006) and OneAPI (GSM Association 2010);
- Definition of the IMS architecture (Knightson et al. 2005) and usage of the SIP protocol (Johnston 2009) as building blocks for a capable NGN (TISPAN 2009);
- Integration of IMS and SDP (Pavlovski 2007).

A service delivery platform (or SDP) is a centralized hub for the creation and integration of all applications and services offered within a network. It is implemented by or on behalf of a network operator, and resides in the "IT" part of the operator's infrastructure.

To avoid standardizing services and facilitate the possibility of a differentiation between operators in terms of something other than pricing, the OMA, and also the ETSI and ITU-T, opted to standardize enablers, called service capabilities by the 3GPP, service support capabilities by the ITU-T and service enablers by the OMA. The SDP allows applications to be abstracted from bearers, channels, enablers, and operational support systems. It goes hand-in-hand with an architectural framework, which describes how it interfaces with the other systems in the operator's infrastructure, including external system. An SDP is used in order to "expose" APIs. They can be opened also to third parties. The reason for adopting the SDP is on the Telco's infrastructure and the possibility to monetize the service exposure, i.e., the offering of APIs, also to third party developers. The SDP is filling a gap in the definition of the IMS at the service layer. IMS service layer is essentially a collection of different application servers (SIP, OSA, CAMEL) not operating in a synergistic way. The goal of SDP is to integrate the different functions of these servers with the management infrastructure of the Telco. To simplify the reasoning behind the equation NGN = IMS + SDP should represent the evolutionary strategy of operators.

2.5.3 *The Peer-to-Peer View on Service Architectures*

Peer-to-peer systems have the goal to optimize the usage of peer resources in order to meet the requirements of a community of users. For instance the Invisible Internet Project (I2P) aims at making the communication between peers secure and anonymized. In order to achieve this goal, the overlay logical network is organized around a few building blocks that can be combined to create an infrastructure on top of which services and applications can be executed guaranteeing anonymity to users (Aked 2011). Any peer will try to create encrypted tunnels toward the peers it wants to communicate with. Tunnels will be outbound and inbounds. The peers needed for creating tunnels are selected by means of a distributed Network database. When the communication has to be created, a lease will be issued so that the inbound and inbound tunnels can be connected and the end parties can communicate. Figure 2.21 illustrates the mechanisms.

Once the network between peers has been constructed, common applications can be launched and used by taking advantage of the new infrastructure.

Another relevant P2P architecture is JXTA (Gradecki 2002), whose goal is to support the creation and the usage of large distributed application using the functionalities offered by specialized peers. JXTA decouples basic functions from services and different applications. At the lower levels, mechanisms for creating pipes

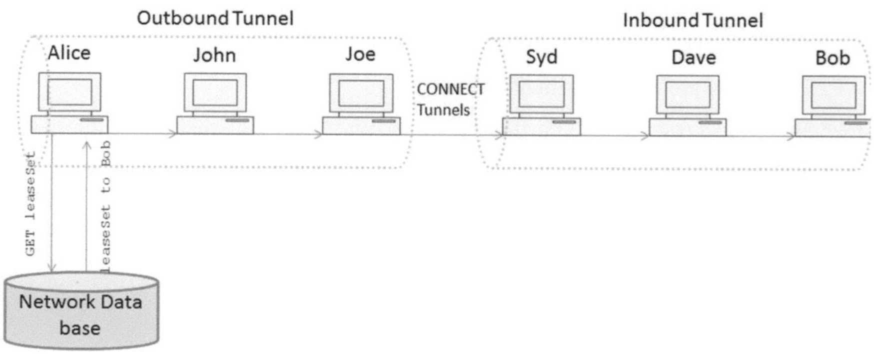

Fig. 2.21 Invisible internet project, I2P

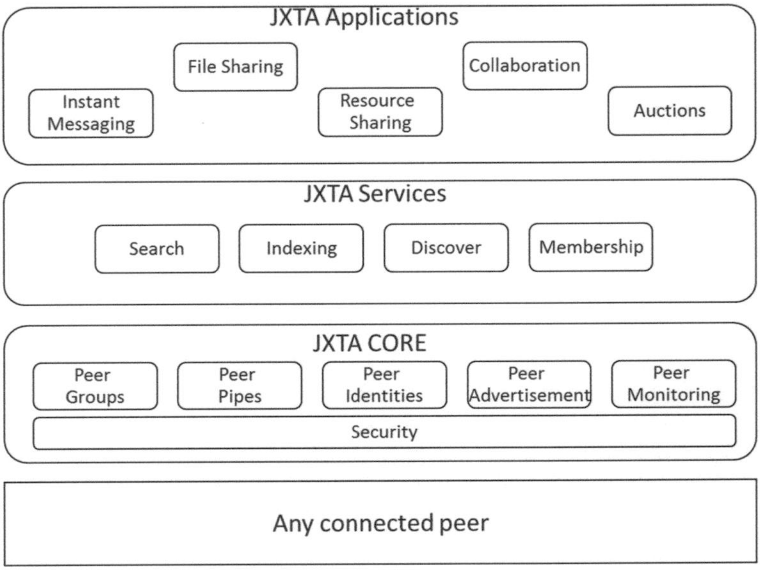

Fig. 2.22 JXTA architecture (Gradecki 2002)

between peers, grouping them, and identifying them are provided. On a higher level, services for discovery of peers, determining the group associated to a specific peer and others are provided. At the upper level, applications are created and used exploiting the JXTA functionalities. Figure 2.22 depicts the architecture.

In spite of their wide use, there are not well established and standardized P2P Platforms, However a few considerations about P2P architectures can be drawn from the previous discussion about architectures

- Complexity of the entire system and simplicity of the single node
  - While nodes are both clients and servers and provide a simple behavior, their aggregation is posing serious issues in terms of management and operation.
- P2P systems are real distributed systems,
  - they need to govern an overlay network (resource naming, addressing, location)
  - the overlay network is totally decoupled from the underlying physical infrastructure. This can introduce inefficiency in how network resources are used.
- Scalability and Robustness
  - P2P systems can scale to millions of nodes
  - Problems in detecting the source of issues in case of malfunctions

- Different ways of programming
  - P2P systems can be implemented as event-driven system but also the communication between two terminal nodes can be client-server.

An important property of P2P systems is their power (Delaney et al. 2001; Manzalini et al. 2010), and in particular their scalability (Hu and Liao 2004; Krishnamurthy et al. 2001), their availability (Bhagwan et al. 2003) and other properties such as stability (Pouwelse et al. 2005). Peer-to-peer networks offer the possibility to aggregate computing, communication and storage capabilities made available by participating peers and to scale up to very large systems. In comparison, a data center requires a lot of investments and planning in order to achieve the desired/needed capacity and scalability.

The total power of a data center is fixed and is given by the number of servers, and related processing and storage capabilities as well as the total bandwidth associated to the data center, in other terms the "power" can be indicated as power (client-cerver cystem) = $\{b_S, s_S, f_S, p_S\}$

where

b_S bandwidth of the server System
s_S storage of the server system
p_S processing in the server system

In a P2P system, the total power is given by the aggregation of individual contributions of peers, so it grows with the number of participants, in a way the "power" could be indicated as Power (P2P System) = $\sum (b_i, s_i, p_i)$

where

b_i bandwidth of node i
s_i storage of node i
p_i processing of node i.

In principle, the aggregated power of a P2P system can grow and compare to large data center when the scale of the system comprises several millions of machines and nodes. However, there are some differences to consider: heterogeneity of the machines, while in a data center there is a high degree of homogeneity (machines are the same) in a P2P system, the aggregating nodes could be very different from each other and also in similar machine the resources (processing, storage, and communications capabilities) could be shared in different ways; the availability of nodes over time, actually each single node can be turned off or leave the P2P network anytime giving a much higher degree of instability to the network. The issue of optimization of using the underlying physical resources has been tackled (V. Gurbani, V. Hilt and I. Rimac) by defining simple protocols for the allocation of resources. Virtualization and self-organization could be other meaningful ways to lessen the problem (Moiso et al. 2010; Manzalini et al. 2010).

A Taxonomy of services and application of P2P (Minerva 2008a, b, c) is depicted in Fig. 2.23.

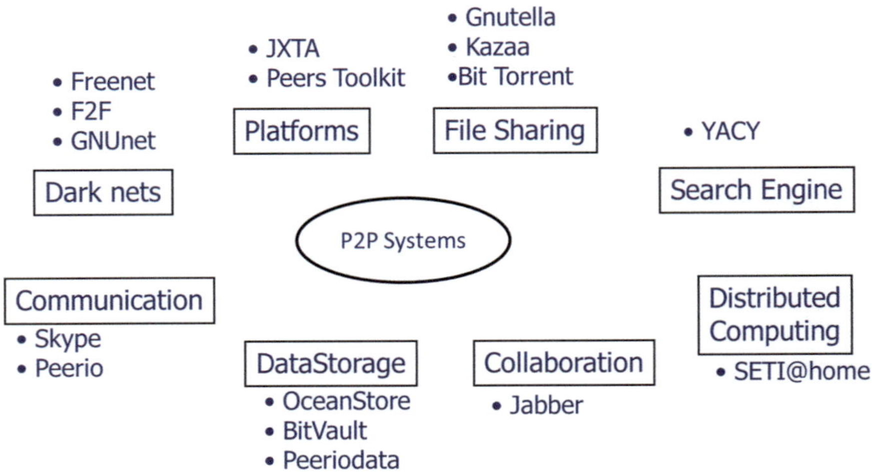

Fig. 2.23 A taxonomy of current P2P services

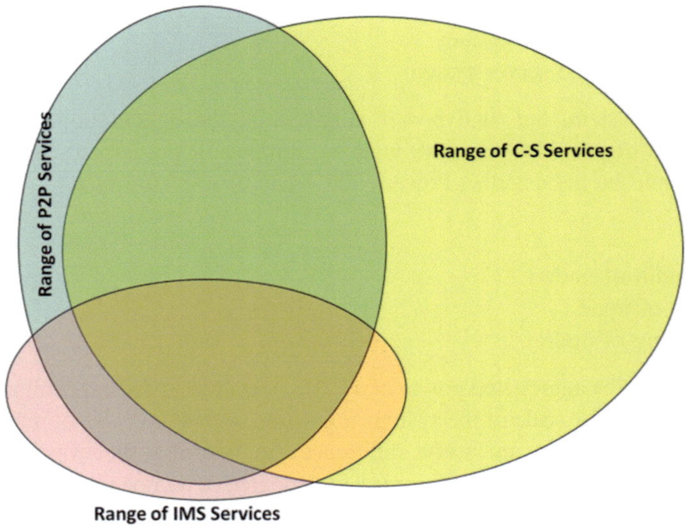

Fig. 2.24 The service span of the considered control paradigm and architectures

The current coverage of services with the three analyzed control paradigm is quite different as depicted in Fig. 2.24. However, P2P seems to be usable for providing services in a widely distributed manner and its potential capabilities are similar to the C-S paradigm, especially if a set of standardized mechanisms could be generally implemented (e.g., distributed hash tables). The value of P2P paradigm is also related to its capability to de-perimetrize services[4] and to disrupt the current status quo. Operators could try to benefit to revert a critical situation in the realm of service offering.

2.6 Findings

This chapter has introduced some of the approaches that are undertaken by major Actors in order to provide services. The distinctive standpoints of C-S, NI, and P2P approaches will be used to identify and discuss some possible scenarios that could occur in the future. It is very significant from a business and scientific perspectives, the increasing interest that is growing around services. The definition of a "service science" comprising several application areas is important because it is paving the way to more precise mechanisms, methodologies, and systems that can fulfill the service requirements. Under this perspective, it should be noted that the user interests and rights are still not fully considered: in fact, many "user-centered" approaches try to understand how to better offer service to the user, and not in guaranteeing to users fair services that deliver value for the right price. This is a deficiency that will be emphasized in the entire document. Chapter 6 will deal with this important aspect of services: i.e., Business Models and the way they are used by Actors of the ICT industry to exploit and take advantage of customers and users. In addition in this chapter, an essential concept related to services is appeared: the more and more proactive role of users. This is a changing factor in the industry because users can play a more relevant role in delivery services and this can have deep social and economic impacts.

Currently, different competing Stakeholders are using Services and Business Models in order to alter, increase or consolidate their positions in the value chain or to change the "status quo" yielding to new opportunities. Examples are Voice over IP (VoIP) or text messaging services like Whatsapp that are used by WebCos for acquiring new users while at the same time disrupt a consolidated market for Telcos. In addition a user-based model could lead to disruptive scenarios in which traditional Stakeholders can be bypassed in favor of service and network infrastructures totally based on user provided resources. The current trend of Fog Computing or Edge Computing (Shanhe et al. 2015) is a step toward these possibilities. Interaction and control paradigms have been introduced; they are important because they affect the way, the services are implemented and provided. By choosing a control paradigm "a priori" without considering the service, the users and their needs can push toward cumbersome solutions instead of focusing clearly

[4]Services in a P2P system can be implemented wherever there is processing and storage power (provided by users), individual peers provide also connectivity that is usable independently from the Operator that is physically supporting it. Services can be deployed and provided by making use of resources that are not necessarily attached to the specific TelCo Network; actually they can be geographical distributed.

on solving problems related to the service delivery. In this way, they are tackling the issue of how a platform supporting a specific paradigm can be used to deliver a specific service. This lack of flexibility in service design and implementation is a major drawback of many platforms that do impose very stringent control means on services and applications. As seen, Network Intelligence is especially prone to this problem because of the tight coupling of network and service functionalities. The C-S and the P2P paradigms are more flexible and can support a variety of possible interaction and control mechanisms reducing this dependency. The expressiveness of a paradigm is of paramount importance in order to create efficiently and effectively new services. Section 2.5 has discussed how the expressive power of a paradigm and its supporting infrastructure has a direct consequence on the possibility to create a large number of interesting services. This leads to the issue of service architectures. In this chapter, the basic service architecture approaches have been considered showing that the C-S one has currently an advantage, while the P2P is more complicated, but it could support in a very flexible and open way many services. The NI paradigm and related architectures seem the one that is less flexible and reusable for providing new services. In the following chapters, the issues related to these architectures will be further elaborated.

Bibliography

Adler RM (1996) Distributed coordination models for client/server computing. IEEE Comput 28(4):14–22

Aked S (2011) An investigation into darknets and the content available via anonymous peer-to-peer file sharing. In: 9th Australian information security management conference. Edith Cowan University, Perth, Australia

Aoyama T (2009) A new generation network: beyond the internet and NGN. IEEE Commun Mag 47(5)

API, wikipedia (2013) Application programming interface. Available at https://en.wikipedia.org/wiki/Application_programming_interface. Wikipedia, last accessed may 2013

Baglietto, P, Maresca, M, Stecca M, Moiso C (2012) Towards a CAPEX-free service delivery platform. In: 16th international conference on intelligence in next generation networks (ICIN). IEEE, Berlin

Berndt H, Darmois E, Dupuy F, Inoue Y, Lapierre M, Minerva R, Minetti R, Mossotto C, Mulder H, Natarajan N et al (1999) The TINA book: a co-operative solution for a competitive world. In: Inoue Y, Lapierre M, Mossotto C (eds) Prentice Hall

Berndt H, Minerva R (1995) Definition of service architecture. Deliverable. TINA Consortium, Red Bank

Bertin E, Crespi N (2013) Architecture and governance for communication services. Wiley-ISTE, London

Bhagwan R, Savage S, Voelker GM (2003) Understanding availability. Peer-to-peer systems II. Springer, pp 256–267

Booth D et al (2004) Web services architecture. W3C working group note, W3C, W3C

Box D et al (2000) Simple object access protocol (SOAP) 1.1. Standard, W3C

Cachin C, Guerraoui R, Rodrigues L (2011) Introduction to reliable and secure distributed programming. Springer, Berlin

Cardoso J, Konrad V, Matthias W (2009) Service engineering for the internet of services. In: Joaquim F, José C (eds) Enterprise information systems - lecture notes in business information processing, vol 19. Springer Lecture Notes in Business Information Processing, Berlin Heidelberg, pp 15–27

Camarillo G, Garcia-Martin MA (2007) The 3G IP multimedia subsystem (IMS): merging the Internet and the cellular worlds, John Wiley and Sons, New York

Chinnici R, Moreau JJ, Ryman A, Weerawarana S (2007) Web services description language (wsdl) version 2.0 part 1: Core language. Recommendation, Boston - Geneva: W3C

Chowdhury NM, Boutaba R (2010) A survey of network virtualization. Comput Netw (Elsevier) 54(5):862–876

Chung PE et al (1998) DCOM and CORBA side by side, step by step, and layer by layer. C++ Rep 10(1):18–29

Clement, L, Hately A, Rogers C, von Riegen T et al (2004) UDDI version 3.0. 2. Specification technical committee draft, UDDI Org

Copeland R (2009) Converging NGN wireline and mobile 3G networks with IMS. CRC Press, London

Curbera F, Leymann F, Storey T, Ferguson D, Weerawarana S (2005) Web services platform architecture: SOAP, WSDL, WS-policy, WS-addressing, WS-BPEL, WS-reliable messaging and more. Prentice Hall PTR, Englewood Cliffs

Curbera F, Duftler M, Khalaf R, Nagy W, M N, Weerawarana S (2002) Unraveling the web services web: an introduction to SOAP, WSDL, and UDDI. IEEE Internet Comput 6(2):86–93

Dabek F, Zeldovich N, Kaashoek F, Mazières D, Morris R (2002) Event-driven programming for robust software. In: Proceedings of the 10th workshop on ACM SIGOPS European workshop. ACM, Saint-Emilion, France, pp 186–189

Datla D et al (2012) Wireless distributed computing: a survey of research challenges. IEEE Commun Mag (ComSoc) 50(1):144–152

Delaney B, Catarci T, Little TDC (2001) The power of P2P. IEEE Multimedia 8(2):100–103

Deutsch, P (1995) Fallacies of distributed computing. White Paper, wikipedia

Di Caprio G, Moiso C (2003) Web services and parlay: an architectural comparison. In: Proceedings of ICIN. ICIN, Bordeaux, France, pp 1–6

Faynberg I, Shah NJ, Gabuzda LR, Kaplan MP (1996) The intelligent network standards: their application to services. McGraw-Hill Professional, New York

Fette I, Melnikov A (2011) The websocket protocol. RFC 6455, W3C, Boston, Geneva

Fielding RT, Taylor RN (2002) Principled design of the modern web architecture. ACM Trans Internet Technol (TOIT) (ACM) 2(2):115–150

Foster I (1995) Designing and building parallel programs. Addison-Wesley Reading, Boston

Foster I, Zhao Y, Raicu I, Lu S (2008) Cloud computing and grid computing 360-degree compared. Grid computing environments workshop. IEEE, Austin, TX, USA, pp 1–10

Gong Y, Ying, Z, Lin M (2012) A survey of cloud computing. In: Proceedings of the 2nd international conference on green communications and networks 2012 (GCN 2012). Springer, Gandia, Spain, pp 79–84

Gradecki JD (2002) Mastering JXTA: building java peer-to-peer applications. Wiley, New York

Grelck C, Scholz SB, Shafarenko A (2010) Asynchronous stream processing with S-Net. Int J Parallel Prog (Springer) 38(1):38–67

GSM Association (2010). Home-3rd party access project-OneAPI. White Paper, GSM Association, London

He W, Xu L (2012) Integration of distributed enterprise applications: a survey. IEEE Trans Industr Inf 99:1

Hickson I (2011) The websocket api. Working draft WD—websockets. W3C, Boston, Geneva

Hu SY, Liao GM (2004) Scalable peer-to-peer networked virtual environment. In Proceedings of 3rd ACM SIGCOMM workshop on Network and system support for games. ACM, Portland, OR, USA, pp 129–133

Isenberg DI (1998) The dawn of the “stupid network”. netWorker, pp 24–31

ITU (2011) ITU and its activities related to internet protocol (IP) networks. http://www.itu.int. In: ITU (ed). 4 Apr 2011. http://www.itu.int/osg/spu/ip/glossary.html. Accessed 26 May 2013
Johnston AB (2009) SIP: understanding the session initiation protocol. Artech House Publishers, London
Jones S (2005) Toward an acceptable definition of service [service-oriented architecture]. IEEE Softw 22(3):87–93
Knightson K, Morita N, Towle T (2005) NGN architecture: generic principles, functional architecture, and implementation. IEEE Commun Mag 43(10):49–56
Koskela T, Kassinen O, Harjula E, Ylianttila M (2013) P2P group management systems: a conceptual analysis. ACM Comput Surv 45(2): 20, 25
Krakowiak S (2003) What is middleware. www.objectweb.com. http://middleware.objectweb.org/. Accessed 26 May 2013
Krieger D, Adler RM (1998) The emergence of distributed component platforms. IEEE Comput 31(1):43–53
Krishnamurthy B, Wang J, Xie Y (2001) Early measurements of a cluster-based architecture for P2P systems. In: Proceedings of the 1st ACM SIGCOMM workshop on internet measurement. ACM, Burlingame, CA, USA, pp 105–109
Kubiatowicz JD (1998) Integrated shared-memory and message-passing communication in the alewife multiprocessor. PhD Thesis, MIT, Boston
Kwon YW, Tilevich E, Cook WR (2011) Which middleware platform should you choose for your next remote service? Serv Oriented Comput Appl (Springer) 5(2): 61–70
Kwon YW, Tilevich E, Cook WR (2010) An assessment of middleware platforms for accessing remote services. In: IEEE international conference on services computing (SCC). IEEE, Miami, FL, USA, pp 482–489
Lankhorst M (2013) Enterprise architecture at work: modelling, communication and analysis. Springer, Berlin
Laplante P, Hoffman RR, Klein G (2007) Antipatterns in the creation of intelligent systems. IEEE Intell Syst 22(1): 91–95
Laurent SS, Johnston J, Dumbill E, Winer D (2001) Programming web services with XML-RPC. O'Reilly Media, Incorporated, Sebastopol, CA, USA
Lenk A, Klems M, Nimis J, Tai S, Sandholm T (2009) What's inside the Cloud? An architectural map of the Cloud landscape. In: Proceedings of the 2009 ICSE workshop on software engineering challenges of cloud computing. IEEE, Vancouver, Canada, pp 23–31
Leopold C (2001) Parallel and distributed computing: a survey of models, paradigms, and approaches. Wiley, New York
Licciardi CA, Minerva R, Cuda A (2000) TINA is dead, long live TINA: toward programmable solutions for next generation services. In: TINA conference. TINA_C, Paris, pp 16–20
Lua EK, Crowcroft J, Pias M, Sharma R, Lim S (2005) A survey and comparison of peer-to-peer overlay network schemes. IEEE Commun Surv Tutorials 7(2):72–93
Maes SH (2010) Next generation telco service providers: Telco 2.0 and beyond. Huawei White Paper, Huawei
Magic, instructional media +. (2012) Application programming interface. http://www.immagic.com/. http://www.immagic.com/eLibrary/ARCHIVES/GENERAL/WIKIPEDI/W120623A.pdf. Accessed 26 May 2013
Maly RJ, Mischke J, Kurtansky P, Stiller B (2003) Comparison of centralized (client-server) and decentralized (peer-to-peer) Networking. Semester thesis, ETH Zurich, Zurich, Switzerland, pp 1–12
Manzalini A, et al (2010) Self-optimized cognitive network of networks. Future Network and Mobile Summit 2010. IEEE, Florence, Italy, pp 1–6
Manzalini A, Minerva R, Moiso C (2010) Exploiting P2P solutions in telecommunication service delivery platforms. In: Antonopoulos N, Exarchakos G, Li M, Liotta A (eds) Handbook of research on P2P and grid systems for service-oriented computing: models, methodologies and applications. Information Science Reference, Hershey, PA, pp 937–955

Manzalini A, Minerva R, Moiso C (2009) If the web is the platform, then what is the SDP? ICIN. IEEE, Bordeaux, pp 1–6

Margara A, Cugola G (2011) Processing flows of information: from data stream to complex event processing. Proceedings of the 5th ACM international conference on distributed event-based system. ACM, New York, pp 359–360

Merwe JVD, Dawoud D, McDonald S (2007) A survey on peer-to-peer key management for mobile ad hoc networks. ACM Comput Surv (CSUR) 39(1). Article n. 1

Metzler J (2011) Virtualization: benefits, challenges, and solutions. White Paper, Riverbed Technology, San Francisco

Minch R (2009) Oracle's e-business suite: an N-tier, networked application. ITM305-003. Boise State University, Boise City, ID, USA

Minerva R (2008a) On some myths about network intelligence. In: Proceedings of international conference on intelligence in networks-ICIN2008. ICIN, Bordeaux, France, pp 1–6

Minerva R (2008b) On the art of creating services: do different paradigms lead to different services? J Telecommun Manag 1:33–45 Henry Stewart Publications

Minerva R (2008c) On the importance of numbers and names for a 4G service architecture. In: Annual review of wireless communication, vol 3. IEC

Minerva R (2004) The death of network intelligence? In: Proceedings of international symposium on services and local access 2004 (ISSLS). Edinburgh

Minerva R, Bell S (2010) Boundary blurring between telecom and the internet. Connect World

Minerva R, Moiso C (2000) Will the "circuits to packets" revolution pave the way to the "protocols to APIs" revolution? CSELT Techn Rep 28(2):213–226

Minerva R, Manzalini A, Moiso C (2011) Towards an expressive, adaptive and resource aware network platform. In: Prasad A, Buford J, Gurbani V (eds) Advances in next generation services and service architectures. River Publisher, pp 43–63

Minerva R, Moiso C, Viviani G (1998) The middleware at the verge between internet and telecom services. CSELT Tech Rep 26:657–672

Moerdijk AJ, Klostermann L (2003) Opening the networks with Parlay/OSA: standards and aspects behind the APIs. IEEE Netw 17(3):58–64

Moiso C et al (2010) Towards a service ecology for pervasive networked environments. Future Network and Mobile Summit 2010. IEEE, Florence, Italy, pp 1–6

Moriana Group (2004) Service delivery platforms and telecom web services—an industry wide perspective. Thought Leader Report, Moriana Group, Egham, Surrey, UK

NGNI Group (2009) FUSECO playground. Report, Fraunhofer Fokus, Berlin

NGNI Group (2005) Open IMS playground. Report, Fraunhofer Fokus, Berlin

NGNI Group (2006) Open SOA Telco playground. Report, Fraunhofer Fokus, Berlin

NGNI Group (2012) Smart communications playground. Report, Fraunhofer, Berlin

NTIA (1996) Telecommunication service. Web page, national telecommunications and information administration. NTIA, Washington, USA

Object Management Group (2006) CORBA Component Model 4.0. Specification formal/06-04-01. OMG, Boston

Ohtsuki HC, Lieberman HE, Nowak AM (2006) A simple rule for the evolution of cooperation on graphs and social networks. Nature 441(7092):502–505

Orfali R, Harkey D, Edwards J (2007) Client/server survival guide. Wiley, New York

Pan J, Paul S, Jain R (2011) A survey of the research on future internet architectures. IEEE Commun Mag 49(7):26–36

Parlay Group (2006) ParlayX 2.1 specification. Specification, The Parlay Group, London, UK

Partridge C, Bailey I (2010) An analysis of services. Report Unclassified, Model Future. UK Ministry of Defence, London, UK

Pavlovski CJ (2007) Service delivery platforms in practice [IP Multimedia Systems (IMS) Infrastructure and Services]. IEEE Communications Magazine (IEEE) 45(3):114–121

Pavlopoulos A, Cooper R (2007) Towards a survey of the deployment space for internet applications. In: BNCOD '07 24th British national conference on databases. Springer, Glasgow, UK, pp 110–119
Peterson L, John W (2007) Overview of the GENI architecture. GENI Design Document GDD-06-11, GENI: Global Environment for Network Innovations, Washington, DC, USA: NSF
Pouwelse, J, Garbacki P, Epema D, Sips H (2005) The bittorrent P2P file-sharing system: measurements and analysis. Peer-to-peer systems. Springer, Berlin, pp 205–216
Prodan R, Ostermann S (2009) A survey and taxonomy of infrastructure as a service and web hosting cloud providers. In: 10th IEEE/ACM international conference on grid computing. IEEE, Banff, Alberta, Canada, pp 17–25
Proper E (2010) Trends in enterprise architecture research. In: Aier S, Ekstedt M, Matthes F, Proper E, Sanz JL (eds) 5th TEAR Workshop. Springer, Delft
Richards R (2006) XML-RPC. In: Proceedings of PHP XML and web services (Apress), pp 595–631
Riede C, Al-Hezmi A, Magedanz T (2008) Session and media signaling for IPTV via IMS. In: Proceedings of the 1st international conference on MOBILe wireless MiddleWARE, operating systems, and applications. ICST (Institute for Computer Sciences, Social-Informatics and Telecommunications Engineering), Brussels. Article n. 20
Ripeanu M, Foster I, Iamnitchi A (2002) Mapping the gnutella network: properties of large-scale peer-to-peer systems and implications for system design. arXiv preprint cs/0209028, arXiv
Rotem-Gal-Oz A (2006) Fallacies of distributed computing explained. http://www.rgoarchitects.com/Files/fallacies.pdf, 20
Sadashiv N, Dilip Kumar SM (2011) Cluster, grid and cloud computing: a detailed comparison. In: 6th international conference on computer science and education (ICCSE). IEEE, Singapore, pp 477–482
Sahoo J, Mohapatra S, Lath R (2010) Virtualization: a survey on concepts, taxonomy and associated security issues. In: Second international conference on computer and network technology (ICCNT). IEEE, Bangkok, Thailand, pp 222–226
Saltzer JH, Reed DP, Clark DD (1984) End-to-end arguments in system design. ACM Trans Comput Syst (TOCS) (ACM) 2(4):277–288
Santoro N (2006) Design and analysis of distributed algorithms, vol 56. Wiley-Interscience, New York
Schaffer HE (2009) X as a service, cloud computing, and the need for good judgment. IT Prof (IEEE) 11(5):4–5
Service Science, Wikipedia (2013) Service science, management and engineering. Wikipedia. http://en.wikipedia.org/wiki/Service_science. Accessed 27 May 2013
Shanhe Y, Li C, Li Q (2015) A survey of fog computing: concepts, applications and issues. Workshop on Mobile Big Data (pp. 37–42). ACM
SOA-Wikipedia (2013) Definition of service oriented architecture. http://en.wikipedia.org/wiki/Service-oriented_architecture. Accessed 26 May 2013
Song J, Chang MY, Lee SS, Joung J (2007) Overview of itu-t ngn qos control. In: IEEE (ed) Commun Mag 45(9):116–123
Sward RE, Boleng J (2012) Service-oriented architecture (SOA) concepts and implementations. In: ACM SIGAda Ada letters. ACM, New York, pp 3–4
Taylor IJ (2005) From P2P to web services and grids: peers in a client/server world. Springer, Berlin
Taylor IJ, Harrison AB (2009) From P2P and grids to services on the web: evolving distributed communities. Springer, Berlin
Thampi, SM (2009) Introduction to distributed systems. preprint arXiv:0911.4395, arXiv preprint arXiv:0911.4395
TISPAN (2009) NGN functional architecture. ETSI ES 282 001. ETSI, Sophia Antipolis
Turner M, Budgen D, Brereton P (2003) Turning software into a service. IEEE Comput 36(10):38–44

Vahdat A (2012) Symbiosis in scale out networking and data management. In: Proceedings of the 2012 international conference on management of data. ACM, Scottsdale, Arizona, USA, pp 579–580

Walkden M et al (2002) Open service access: advantages and opportunities in service provisioning on 3G mobile networks definition and solution of proposed Parlay/OSA specification issues. OSA Specification issues, Project P1110 Technical Information EDIN. Eurescom, Heidelberg, Germany

Yagi, H, Ohtsu K, Wakano M, Kobayashi H, Minerva R (1995) TINA-C service components. In: Proceedings of TINA95 integrating telecommunications and distributed computing-from concept to reality. TINA_C, Melbourne, Australia

Znaty S, Hubaux JP (1997) Telecommunications services engineering: principles, architectures. ECOOP Workshops. Springer, Jyväskylä, Finland, pp 3–11

Chapter 3
Technological Evolution of the ICT Sector

We have not reached the cusp of technology evolution yet. Fresh innovations are lined up for disrupting established “mainstream” technologies. Exploring technology evolution provides clues to the next wave of transformation and strategic decisions. Data centers technologies demonstrate how service “factories” can merge client–server front end, with an elaborate distributed computing and cloud behind, but also illustrate that the technological gap between Webco and Telco offerings is increasing due to web agility and DIY attitudes.

Virtualization has far reaching impact, not only on efficiency, but on business opportunities—XaaS business models are all heavily based on it. Distributed applications, cloud, and virtualization mechanisms allow Telcos room for de-parameterisation and globalization.

Big data facilitates services never conceived before, combining physical world and virtual world activities. While operators’ massive data can now be put to use, there is also intensified network deployment of sensors and machine-to-machine communication, with emerging new breeds of applications. Wearable devices will bring a wealth of new functionalities that are powered by the growing terminal capabilities, and this will revolutionize the service environment in return. Hence new technologies disrupt the old, and create demand for the new. Technology is the clue.

3.1 Introduction

The goal of this chapter is to represent the current trends and developments characterizing the evolution of the ICT industry. The rationale behind this is to look at how the industry is progressing from the technological point of view. In fact the evolution of some basic technologies like processing, storage, sensing, and others will be briefly analyzed in order to project it in a time frame of ten years. This should provide a vision of what will be possible at the technological level. In addition,

R. Minerva and N. Crespi, *Networks and New Services: A Complete Story*,
Internet of Things, DOI 10.1007/978-3-319-33995-5_3

the evolution of some relevant software mainstream technologies will be considered aiming at evaluating how they will shape the future. This chapter will provide a sort of common ground on top of which to reason about how and why to use available technologies and solutions. On top of this, some initial considerations will be put forward in order to identify some inflection points (i.e., some points in time in which the technical evolution can enable new services and solutions that where not implementable or deployable up to a few years ago: e.g., pervasiveness of networks and devices). They are not singularity points, but these "inflections" could identify changes in the linear evolution and can anticipate the singularities that can enable new services and new technologies and some possible disruptions in this "mainstream" approach. Actually this chapter tries to analyze the major technologies forming the mainstream and to identify whether they could be superseded by better ones on the basis of possible evolution of enabling technologies.

Under this respect the evolution of data center, virtualization, and cloud computing seem to represent the most likely evolution toward the creation and consolidation of large systems capable of serving millions of users by means of the C-S paradigm. The prevailing vision is the one that foresees the concentration of services in "hyper" clouds made out of millions of servers. The technologies supporting this concentration are consolidating today and services will be moved to the cloud.

This move of services and applications in the cloud comes with a manifest consequence: large quantities of data will be gathered, collected, and processed within the cloud. In addition, new data stemming from the integration of different data sets will be created in these massive data centers. Almost any action of users could be observed, screened, and recorded. This big data approach could be pursued and supported in different ways in various parts of the world. Europe seems to adopt an approach respectful of the privacy of the users, but this attitude may have very heavy consequences on an industrial level (i.e., giving an advantage to the US industry). A user centered approach in (personal) data management could be one of the inflection points.

Finally, this chapter will briefly analyze the evolution of (mobile) terminals. The reason for this is that new terminals are very capable in terms of processing, storage, communications, and sensing. Their power will grow also in the future making them an essential factor in the evolution of the services. Actually personal terminals could be seen as service platforms capable of delivering many important functions to the users.

3.2 The ICT Ecosystem

The technological evolution of the ICT sector is confirming its pace of developments: the Moore Law [i.e., the number of transistors on integrated circuits doubles approximately every 18 months (Schaller 1997)], the Kryder Law [i.e., magnetic disk areal storage density increases at a pace much faster than the doubling transistor count occurring every 18 months in Moore's Law (Walter 2005)], the Nielsen's Law [stating that network connection speeds for high-end home users would increase 50 % per year, or double every 21 months (Nielsen 1998)] and the Koomey Law (Koomey 2010) (i.e., the number of computations per joule of energy dissipated has been doubling approximately every 1.57 years) are holding valid. There is not an acceleration but a constant growth, nevertheless processing, storage, and communications are reaching capabilities that can change the way in which ICT is used and perceived by users. Even if the ICT industry is not reaching singularity points,[1] it is approaching situations in which the software technical solutions can substitute hardware-based ones. These can be considered inflection points, i.e., a context in which a technical solution is viable thanks to the increased processing, storage, and communications capabilities. These inflection points anticipate singularity points and will determine and influence the technological evolution of the ICT industry. Some main considerations are worth at this point:

1. General-purpose hardware will be more and more capable (in the sense that it will provide more processing power combined with the possibility to store more information) and will progressively substitute specialized one.
2. The power consumption of hardware will be optimized and will bring benefits to both big data centers and portable user equipment.
3. Software will be the distinctive factor of any device, service, and system.

So far, no laws are governing the technological evolution of sensors and/or actuators, but it is likely that sensing and actuation capabilities will benefit from the general evolution of processing, storage, and communication capabilities. Determining how much they will progress still needs to be determined. However, also this technological branch will influence the evolution of ICT. Some forecasts discussed in Zaslavsky et al. (2013) put in front the hypothesis of having up to 100 billion of sensors. This means that thousands of sensors could surround each human. It is not clear when this situation will be common.

Keeping an outlook to 2020 and beyond, the forecasts depicted in Table 3.1 can be considered (following the current pace of evolution) for some interesting technologies in the ICT sector.

So the environment in which users will be embedded will be visual, highly interactive, and responsive. ICT capabilities will be taken for granted and embedded communication will promote the idea of free connectivity to customers. On the other hand, the progress of local area communications and direct communication in the terminals will enforce this capability.

Terminals will be a differentiated factor in the provision of services and they will drive the developments in terms of services and applications. In this sense, a

[1]i.e., a singularity point in this context is a point in time in which the speed and the results of technological evolution will cause disruptive effets on the current ICT sector and in particular, in Telecommunications leading to a complete transformation of its models, markets and approaches.

Table 3.1 Technology roadmaps

Technology sector	Advancements and perspective to 2020 and beyond
Processing	• Capacity doubles every 18 months • Reduction in power consumption • Considerable cost reduction • Possibilities: – Multicore systems to the general public – Use of ARM-based systems also in Data Centers • Development of non-VonNeumann architectures
Storage	• Capacity doubles every 18 months • Cost reduction (over 375.000 times over initial developments) • Possibilities: – Flash memories with 1–5 Tb – Hard Drives with tens of TBs – Integration of flash memory with wireless technologies – New technologies such as memristor (could help in decreasing the space needed for electronic components increasing their processing speed)
Fiber	• The progress in optoelectronics and the possible decrease in price due to production volumes will make fiber-based solution convenient. This will be emphasized by the possibilities to: – Multiplication of the available Lamba channels – Increase the capacity of the single fiber (up to 100 Tb/s) – New technological advances (e.g., attoseconds optics)
Radio	• Radio technologies are close to reach the theoretical limits of the spectrum, this implies – The need to reduce the radius of Cells (higher distribution of devices) – Reuse and allocation of new portion of spectrum (e.g., 700 MHz) – The introduction of new techniques for interference resolution (Cognitive Radio) – The usage of new technologies such as: lowPower and nanoimpulse systems or OAM (orbital angular momentum) to cram theoretically infinite transmission channels per given frequency (in lab 2.5 Tbs)
Local Networks	• Local connectivity on the increase especially for WiFi based solutions with speed that by the end of decade will exceed 1Gbs • Embedded communication in consumer electronics and household appliances and possible also in car and vehicles • Mesh networking and direct communication capable of supporting applications and services
Visualization	• The evolution of screen technologies (e.g., 8 K screens) will need communication channels over 180 and up to 500 Mbps • Screens of eBooks will have a definition comparable to paper and will be thin and flexible • Micro projection will be available in terminals • Screen with embedded camera (e.g., Apple patent) • Wide deployment of embedded screens in the environments (e.g., tag price will be substituted by screens) • Semitransparent Screens for windshields and glasses • Gesture and natural interaction with screens

(continued)

Table 3.1 (continued)

Technology sector	Advancements and perspective to 2020 and beyond
Sensors	• Billions of sensors deployed and activated by the end of the decade • Sensors as native parts of products • Virtual Sensing (i.e., extraction of information from sensor data)
Terminals	• Terminals will benefit from the pace of technical evolution in terms of processing, storage and communication capabilities • Terminals will be a service platform in the hands of users • Always best-connected strategies will be implemented by the terminal and not by the networks • FON like networking will be used especially in denser populated areas • Open environments will compete with walled garden solutions • New interactive terminals will appear (goggles, watches and the like)

competition between closed and open environments will be going on with consequences on the share of the application market. Open environments repropose the richness of the web, closed systems ensure better users' experience and possibly higher degree of control and security. However, any winning technology will be related to terminals in a way or another (because it will run on terminals or because it will enable new services for terminals).

In the rest of this chapter, more details on specific aspects of the technological evolutions will be briefly discussed.

3.3 Data Center Evolution

Data centers are one of the battlefields for the future of communication market. Actually they are for many global providers, one of the two horns for the market strategy. The application market is dominated by the client–server paradigm, so a strong presence on the server side or having a prominent market share on it is fundamental. For this reason, many web companies do not consider data centers as cost center. They are instead considered innovation points. In fact companies like Google put a substantial effort and investments on the development of such infrastructure (Platform; Barroso et al. 2003; Dean and Barroso 2013; Minerva and Demaria 2006). The technological evolution of data centers is dominated by some issues that owners try to solve

- Power consumption and heat production;
- Replication, deduplication, distribution, and impacts of communication costs;
- Increase and use of processing and storage;
- Programmability and leverage of highly distributed systems (Ghemawat et al. 2003; Pike et al. 2005; Burrows 2006; Chang et al. 2008).

These issues are intertwined and determine many interesting approaches carried out by the data center owners: for instance the need to reduce power consumptions and reduce space allocation is pushing toward the usage of ARM (Advanced RISK Machine) based architectures (see for instance Dell and HP-related announcements,[2,3]). In addition, the need of controlling how connectivity between different data center is used has pushed toward the usage of novel communications solutions that move to communications infrastructure. Some concepts have already been applied with success in the processing and storage: i.e., virtualization and use of low cost general-purpose hardware (Open Networking Foundation 2012). It is not a case that Google is a pioneer of the software-defined networking (SDN) with his own solution (Vahdat 2012), Amazon is following this route as well and its research and experiments in this area are consolidating[4] (Miller et al. 2011). Facebook is active in this area (see its participation in Open Networking Foundation[5]) and it is experimenting on a small-scale system[6] the benefits of flexible allocation of networking resources [now termed as software-defined networking, SDN (McKeown 2008)].

The strive for reaching new levels of distribution in processing and storage capabilities led to interesting techniques in data management [from Google's Big Table (Chang et al. 2008) to NoSQL databases (Leavitt 2010; Stonebraker 2010) up to Amazon's Dynamo (DeCandia et al. 2007), and the like], in parallel computing [e.g., Google's Map-Reduce (Dean and Ghemawat 2010), Hadoop (White 2010), and others].

Facebook approach is more conservative and it aims at the definition and possible standardization of a viable general purpose-based data center infrastructure, the Open Compute Project (Hsu and Mulay 2011), to be defined with the support of the community and other web companies.

In trying to solve their problems, the Web companies have taken the approach to ensure an answer to the "clients" in the shortest possible time (e.g., search queries). In terms of the consistency, availability, and partition tolerance (CAP) theorem (Brewer 2000, 2012), these companies (e.g., Google and Amazon) have chosen to enforce distribution of functions and availability instead that guaranteeing consistency of data. This is a quite new approach that can be acceptable by final customers of search engines, and social networks. This approach has shaped the architectures of the infrastructures of Google, Amazon, and Facebook.

---

[2]http://slashdot.org/topic/datacenter/dell-calxeda-develop-second-arm-server/. Last accessed on April 9th, 2013.

[3]http://www.crn.com/news/data-center/231902061/hp-intros-low-energy-data-center-platform-based-on-arm-servers.htm?itc=xbodyjk. Last accessed on April 9th, 2013.

[4]http://billstarnaud.blogspot.it/2012/02/amazons-simple-workflow-and-software.html. Last accessed April 9th 2013.

[5]See Open Networking Foundation at www.opennetworking.org. Last accessed 30th May 2013.

[6]http://gigaom.com/2013/01/31/facebook-experiments-with-small-scale-software-defined-networking/. Last accessed April 9th, 2013.

Even Twitter's newer architecture is adopting similar approaches[7] (Krishnamurthy et al. 2008; Mathioudakis and Koudas 2010). Actually Twitter is a very interesting architecture because it is pushing for a quasi real-time approach in retrieving data and derives information by them to be exposed to end customers.

For a general discussion about the Google architecture, see (Barroso et al. 2003; Ghemawat et al. 2003; Minerva and Demaria 2006).

Summarizing, the data center approach can be characterized by

- The adoption of disruptive solutions in their design to achieve efficiency, reliability, and scalability, by means of modularity and "self-similarity";
- The adoption of a Do It Yourself, (DIY) attitude, almost the majority of big data center owners have built their own infrastructures (including buildings) using cutting edge technologies (from cooling to power distribution, from servers' architecture to networking protocols) and positioning their infrastructure as a means to cope with and differentiate from competitors;
- Leveraging any progress in processing and storage technology (e.g., ARM-based clusters);
- The mastering of software. Actually the Web companies have mastered their own platforms, starting from general-purpose hardware and introducing innovation in networking, software, and communications by also adopting open source software. They are capable of developing their own software solutions and bringing them to the market for large scale use;
- The development and adoption of new programming paradigms [e.g., MapReduce, Pregel (Malewicz et al. 2010), etc.].

Data centers can be seen as the sets of "centralized" servers, but actually behind a client–server-based front end, they are the epitome of distributed computing. In fact many technologies have been developed, tested, and tuned up into data centers (e.g., big table, chubby, dynamo, and other distributed solutions). So they have a sort of double face: a traditional one in terms of servers receiving and treating requests form a multitude of clients, and a distributed one in which parallel computation, distribution of data, and the newest distributed computing technologies are invented, tried, and nurtured (see Fig. 3.1). The data centers are factories of innovation (when a DIY approach is used). For this reason, adopting the classical Telecom operator approach to data center (the one of intelligent buyer) is not a winning one, in fact the most of innovation is created within DIY data center and is usable after some time in commercial solutions for data center. For this reason, the technological gap between Web companies and Telecom providers in this sector will increase over time. Telcos will be lower cost provider of regional solutions because their solutions will be the one offered by the market and the only competitive asset for differentiation will be the price.

[7]See for instance http://engineering.twitter.com/2010/07/cassandra-at-twitter-today.html for the usage of a nosql solution in Twitter. Last accessed April 9th, 2013.

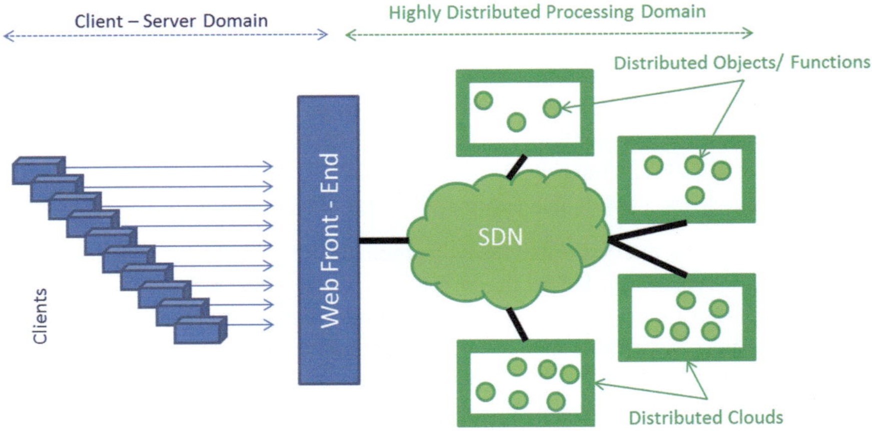

Fig. 3.1 The double face of data centers

On the other side, the data center winning players are moving toward terminals. Each one of them is addressing this market and this technological area. The attempt is clear: to dominate the two aggregation points of intelligence in the Internet: the client and the server side. By doing this, the Web companies are besieging Telecom operators reducing them to a mere commodity and accruing the problems of investment on the communication platform.

3.4 Resource Virtualization

A lot of effort has been spent and devoted to the possibility of virtualize systems and applications on a general-purpose infrastructure. The increasing processing power and storage capabilities help in adopting more and more virtualization technologies within data centers and in distributed systems.

In a quest for flexibility in resource allocation, virtualization techniques are playing an important role. Virtualization "means to create a virtual version of a device or resource, such as a server, storage device, network or even an operating system where the framework divides the resource into one or more execution environments.[8]" Virtualization can be applied at different levels, e.g., hardware, operating system, and application/service. Figure 3.2 represents different perspectives of virtualization.

[8]For a definition of Virtualization see http://www.webopedia.com/TERM/V/virtualization.html.

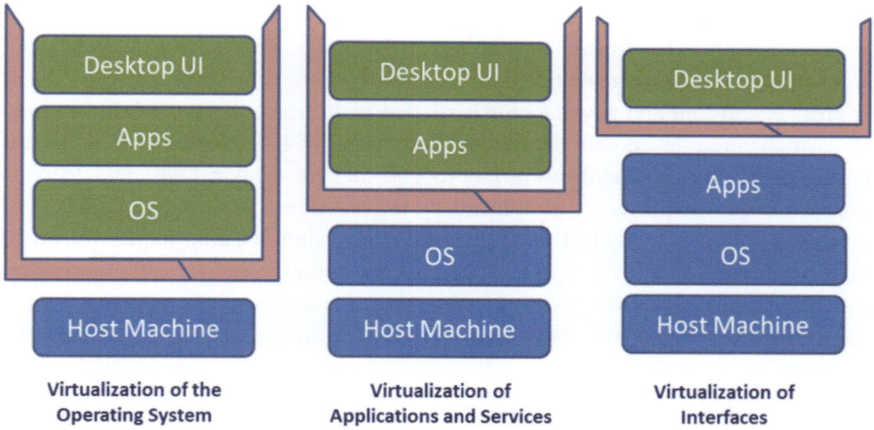

Fig. 3.2 Examples of virtualization at different layers

Generally speaking, Virtualization allows to

- expand hardware capabilities, allowing each single machine to do more simultaneous work;
- control costs and to simplify management through consolidation of servers;
- control large multiprocessor and cluster installations, for example in server farms;
- improve security, reliability, and device independence by using hypervisor architectures;
- run complex, OS-dependent applications in different hardware or OS environments.

These possibilities are quite important and are enabling a new kind of offering: the XaaS (everything as a Service) paradigm is heavily based on the virtualization techniques. In the data center area, virtualization is used to support a whole new approach for the organization of software: every software application can be virtualized in the network and being offered to customers as an on-demand service. New businesses have emerged and also some problems. Some of these issues are strongly related to connectivity: for instance what happens if the data center supporting business critical applications is perfectly running, but there is no connectivity to it? Virtualization, connectivity and data center structure are seen by the customer as a whole, but these features can be sold and operated independently from each other. Another issue is related to the software instances: with traditional software, the customer can use the software even if the producer of it has gone bankrupt, or has not updated the application for a while. With XaaS, if the Provider of the application is not properly running the environment, the customer may not have access to the critical functionalities. This problem could be even exacerbated if the providers go out of business abruptly without giving the time to the customer to migrate to other platforms. One solution could be to run virtualized instances of the application within the domain of the customer.

Another issue to be considered in virtualization is that even if the hardware progress is constant and new capable hardware platforms will succeed during time, the virtualization is another additional layer that will consume processing and storage capabilities. Virtualization has to consider the May's Law[9] and to compensate for the inefficiencies introduced by the software.

In spite of this, in many ICT areas virtualization is already being successfully used and it comes with a lot of disruption as well as new opportunities for creating services and applications. In the Network Control Area, for instance, the network virtualization function initiative, sponsored by ETSI (European Telecommunications Standards Institute) (NFV 2012) is using virtualization for promoting higher levels of functional aggregation and organization of the network functions.

3.5 Distributed Applications and Technologies

Currently, distributed processing is seen as strongly related to cloud computing (Foster et al. 2008; Di Costanzo et al. 2009), this is due to two reasons: for the possibility to create within a single cloud a large distributed system (e.g., Google, Amazon, Twitter) and for the possibility to create large distributed applications on the intercloud (Buyya et al. 2010). However, cloud computing is deployed and used quite differently by several actors. For instance some prominent Web companies use these technologies in order to leverage the infrastructure needed to carry out their principal business (e.g., market place for Amazon and search/advertisement for Google). Their solution of cloud have stemmed as a means to leverage the idle time of their powerful data centers. Many operators have seen a potential revenue generator in this kind of offering and they have put in place cloud computing solutions. The major difference between the operators approach and the Web companies one is related to a recurring concept in the deployment of ICT solutions: deperimeterization of services, i.e., the capability to provide services that are globally accessible and independent from a specific network infrastructure. Typically the services offered by a web company like www.WebCompany.com are accessible to all the users of the web, while services provided by a Telco are available only to the subscribers of the specific service on the specific network. In few words, the Web companies (especially the big ones) have a global footprint, i.e., their services are well known worldwide and customers that use the Internet can typically make use of their offering. Simply put, the Web companies can leverage a long tail effect on the global market, i.e., services provided can be globally accessed by users independently from the geographical location of the users. This allows the Webcos to make small profits

[9]Software efficiency halves every 18 months, compensating Moore's Law (http://en.wikipedia.org/wiki/David_May_%28computer_scientist%29).

in many locations and to reduce the idle time of their infrastructure. Deperimeterization in this case means that worldwide users can access the Web company cloud offering irrespectively of the location of the user. The Telecoms proposition is somehow different: even if their offering could be somehow deperimeterized, these companies do not have a global footprint (even the biggest ones are Regional—China Mobile—or international). As a matter of fact their business is available to their subscribers, bounded to be regional and confined to locations where the specific Telco has a network infrastructure deployed. As discussed in Minerva et al. (2013), in this area the approach of Telco has to be totally different and even more technical challenging compared to Web companies. Telcos have to emphasize their regionalism trying to work out an offering (coupled with a technology platform) that is customizable for the specific regional needs of the customer, in addition, the solution should be supporting some levels of interoperability with other regional solution possibly not owned by the same Telco. In this context the concept of federation of clouds is of the paramount importance. This means to provide a regional solution that is adhering to laws and practices of that specific location (together to the management of the platform) but open to integration with other clouds of resources. Federation is a difficult technical topic that comprises not only IT and software issues, but also communications. In fact regional-federated clouds need to have a good level of interconnection in order to allow applications and data to flow. In other terms Telcos have to try to overcome their regionalism by federating different clouds possibly pertaining to different providers. Web companies cannot count on the local presence, but they can provide "normalized" and cheap solutions (costs are kept down because of the big volumes) that are accessible globally. More considerations and proposal for a taxonomy of cloud solution can be found in the already mentioned paper (Minerva et al. 2013).

However, distributed computing is not and will not be confined to cloud computing. In fact, there are a number of different technologies under this technological area. One first observation is that generally speaking, cloud computing is disregarding the issues related to communication. It is suggested to have a good level of service level agreements in order to cope with issues and problems in connectivity. This is not correct and actually even the first specifications of grid were trying to cope with the network organization for adequately support the processing and storage capabilities (Baroncelli et al. 2005). This approach has led to specifications related to the possibility to control by means of managers the connection establishment, update and tear down. This trend has then been adopted by the OpenFlow (McKeown 2008) initiative and has led to a new revamp of the concept of programmable networks: the software defined networks (Lantz et al. 2010; Gude et al. 2008). Similar considerations have driven Web companies to use these technologies for the optimization of their connectivity capabilities, for instance the G-Scale solution developed by Google (Vahdat 2012).

However, the evolution of highly distributed systems is approaching a new cornerstone: i.e., the integration of social-driven connectivity with small range connectivity can bring to the massive adoption of new mechanisms. In fact, the proliferation of capable terminals[10] will permit to exploit the real capabilities of distributed processing. Terminals will be nodes of a distributed environment and they will be able to determine how to connect each other, which functions to execute locally and how and where to store the data. P2P technologies are a clear demonstration of the capability to create huge infrastructures capable of supporting different classes of services. As discussed in Minerva et al. (2011), the Peer-to-Peer technology (compared to network intelligence and client–server) is the one that can support many computation paradigms (event based, tuple space, or other) so that different applications classes can be supported. This flexibility requires the introduction of some functions (with respect to the easiest paradigm: the client–server) in order to create and support the overlay network functions needed by the nodes for correctly operating and benefit of the distributed infrastructure. This paradigm has been proposed for the creation of global service oriented platforms. For example the Nanodatacenter[11] project (Laoutaris et al. 2008; Valancius et al. 2009) was defined in order to use P2P and BitTorrent (Izal et al. 2004) for the distribution of data in a distributed network formed by access gateways and access points. In this case an operator could try to exploit processing, storage, and communications capabilities of those types of devices already deployed in the customers' homes.

These edge distributed systems will be characterized by the large number of devices connected and by the high dynamicity of the availability of a single node. Actually nodes can come and go in the infrastructure bringing in and out their processing, storage, communications, and sensing capabilities. These systems will not be managed with traditional mechanisms. There is a stringent need to have self-organizing capabilities that have to entirely substitute and make it useless the human intervention. The dynamic of entering in these networks will be so short that no human configuration of the terminals could be fast enough to allow the integration of the nodes. Autonomic capabilities are a need; they are essential to ensure the existence of these types of infrastructures. Due to the spontaneous nature of these aggregations of resources, there is a need to understand how to promote the resources sharing and avoid as much as possible opportunistic behaviors. Studies in social networking, e.g. (Ohtsuki et al. 2006), show that when the benefits over the costs of sharing exceed a parameter k (representing the number of neighbors, or links to them), i.e., $B/C > k$, then an altruistic behavior emerges in communities. In order to reach this situation and maintain a value for the major part of participating nodes, some strategies could be implemented based on game theory in order to increase the global value for all and to optimize the benefits for the single nodes. It is clear that this aggregation of nodes will act and behave as complex systems and programming their behavior will require a lot of innovation and research.

[10]i.e., terminals that have a considerable processing power and storage as well as the ability to use different of connectivity means (e.g., 3 or 4 G, WiFi, Bluetooth).

[11]See http://www.nanodatacenters.eu. Last accessed May 30th 2013.

When the technology will reach this point, a sort of very powerful and highly pervasive platform will be available. In this situation, applications will be organized as dynamic "coalitions" of cooperating service components with these features:

- deployed on dynamic and pervasive "cloud" of computing resources (clusters of servers, users' devices, sensors, etc.)
- provided by multiple actors (users, service providers, enterprises, equipment providers, sensor network providers, etc.)
- "ad hoc" assembled and adapted (according to situated needs, component and resource availability, and their changes).

These coalitions will be the framework for building smart environments that enable new business opportunities beyond traditional value chains. A possible effect is that this new organization can contrast "oligopolies" in the service delivery market. These innovations came from a bottom–up approach and leverages the end-user capabilities. For this reason, a similar trend is difficult to spot and possibly it could emerge as a "change of phase," i.e., when the number of terminals and communication capable devices will reach a critical level, this new way of interacting will suddenly emerge bringing disruption to the communication status quo. In addition cognitive and software-defined radio technologies could give a boost to this approach making it a sort of pervasive collaborative platform.

3.6 Harvesting Data: Big Data and Personal Data

Technology advancements in several sectors (devices, sensors, cloud computing, pervasive connectivity, online services, process automation, etc.) extend the possibility to collect, process, and analyze data. So far this information is scattered and mirrored in different silos pertaining to different companies. These data, however, provide hints and clues about the behavior of users and organization (and things associated to humans) with respect to:

- Activities carried out in the physical world and the virtual worlds
- The activities performed inside the organizations
- The "digital footprint" of people, i.e., the collection of data and logs that describs the interaction of users with digital services.

By nature, "Big Data" is characterized by high volumes, velocity, and variety, requiring scalability for managing and analyzing them. Aiming at supporting the 3 V properties, many technologies have been developed such as

- NoSQL Databases (Leavitt 2010; Cattell 2011);
- Map-Reduce processing framework (Dean and Ghemawat 2008, 2010);
- Real-time data stream processing (Stonebraker et al. 2005) [e.g., Twitter (Bifet and Frank 2010)];
- Semantic representations (Auer et al. 2007);

- Improved data mining for analytics (Piatetsky-Shapiro 2007);
- Elastic resource platforms (Marshall et al. 2010).

The current trend for many companies is being a hub for big data collection in order to exploit the information gathered while conducting business with customers. For instance Telcos are eager to collect and possibly exploit the information collected about the users (surfing habits, connectivity related data and the like). Collecting personal information and profiling people is a promising business. This aggregation role enables new business opportunities because of the possibility to analyze data patterns in order to derive new information, the possibility to mashup different sources of data and information and the possibility to distribute this information to interested customers. Actually data availability and mashup capabilities enable an ecosystem of applications bringing benefits to people, public, and private organizations. Many future applications will leverage these features in order to provide meaningful services to people and organizations. On a more scientific side, the data collection enables the data-intensive scientific discovery [as proposed in the Fourth Paradigm (Hey et al. 2009, 2011)]. Many new scientific discoveries will emerge by the capability to access a large database of differentiated data. For instance collateral effects of combined therapies and mix of medicines are hard to determine in laboratories. Accessing health records of people being cured with these combinations could result in determining if there are counter indications for certain pathologies of groups of patients. Data analysis and crosschecking in fact gives new info and new models. For instance in González et al. (2008), some scientists were able to crosscheck the mobility patterns of people and determining their habits.

Operators have plenty of useful data: they range from the data call record that logs all the calls from and to a user, to location information about a user, to surfing habits, and websites visit of users. These datasets could be used in order to determine and improve the understanding of social interactions or the goods flows within a city. However, these data have to be dealt with in respect to the ownership of them and the privacy of users. A user-centric-based ownership model should be enforced by law and regulations. In this way, users could decide how to leverage their own data by accessing "free applications" in exchange for their data or to try to leverage personal data in other ways (Moiso and Minerva 2012; Minerva and Crespi 2011).

The World Economic Forum (2011) sees the area of personal data as the new oil for the digital market. However, user ownership should be enforced and guaranteed. In this context an approach as the one promoted in Moiso and Minerva (2012) seems to be promising in terms of potential business exploitation, but still very respectful of the user ownership. Simply put, data should be collected and stored into banks of data. Users will then decide if they just want to store them for personal advantage (retrieving and controlling the interaction in the web, retrieve a specific transaction or a document and the like) or to exploit their value in a controlled market. So users can establish policies for using personal data and open up accordingly the access to third parties.

Another trend will emerge: a sort of reverse engineering, i.e., even if data are anonymized, tracking capabilities can be put in place in order to determine who the producer of that data was or whose the data was referring to. Privacy and security will be an integral part of the offering of personal data stores.

3.7 Disruptions in Identity Management

Digital Identity is another issue that has the potential to disrupt the current technical and economical evolution of the Internet. As stated by Clark et al. (2005), the identity management is a tussle more of a technological problems. The companies that can manage the identities of users are in the position to infer and derive the actions and the behaviors of users in their digital interaction within the cyberspace (Landau and Moore 2011). In fact identity management is strongly related to profiling of users, the link between them is an enabler for relating actions and behaviors to the individual. The issue of identity is also related to the rights of the citizen and/or customer: the Internet is global by nature and data "flow" where it is easier to deal with them. There is a great difference in the law enforcement between Europe and the USA in privacy issues. In Europe privacy is considered an important right of the Citizen, in USA privacy is seen as a right of the customer. The consequences of these two divergent approaches are very profound: personal data and identity management in Europe are so regulated that many services (e.g., the study of flows in a smart city) are not possible unless all the involved parties give the consensus to the treatment of data (e.g., the call data record can be used for a service if and only if all the involved parties, i.e., the caller and the callee) agree to its usage. In the USA it is enough to give a general disclaim in the service offering and personal data can be used. As a consequence, the American companies are in a more favorable position to offer these kinds of services over the European ones. In addition the collection and aggregation of data are performed out of Europe in order to avoid this more stringent regulation. The consequence on competition is evident: the US companies can do more than European ones. There is a need to enforce the regulation on all the data of European citizens, but that is very tricky and difficult.

There is a basic observation that should drive the discussion about identity. Who is the owner of the personal identity? In Minerva and Crespi (2011), this issue has been tackled with a disruptive approach in mind. If the owner of the identity (and the related identifiers) is the user, than a great change could take place in the communications and Internet sectors. Just to name two examples: if the phone number is property of the user, then roaming and number portability should be radically changed under the concept that the user identity can be dynamically associated for a small period of time to another network or to a communication environment. Same thing for instance in the Internet, if the email address is not owned by the provider but by the user, then email addresses should be portable from a system to another. This could change the long established market of email. Identity is also intertwined with security. Users should be able to decide how their

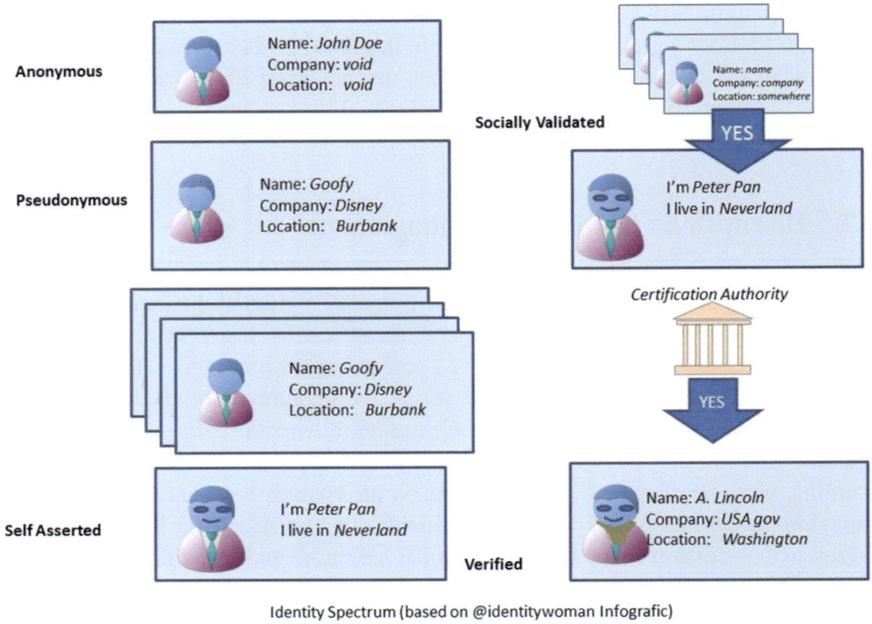

Fig. 3.3 The different modes that constitute the identity spectrum

identity should be dealt with; in www.identitywoman.net a spectrum of identity modes is given (see Fig. 3.3).[12]

User should have the right to be anonymous without any provider trying to impose registration to site or identification of the person; other times the user should be let free to use pseudonymous because there is more than a digital identity associated to an individual; the user should also be capable of self-asserting its identity by means of statement without any authority trying to validate those or requesting proof of statements; a user could also be identified by its friends and then being socially validated; the usual case of validation by a third party should be the exception rather than the norm (as it is today) and this form of identification should be used in limited circumstances (e.g., when using home banking applications). This identity spectrum shows how users are spoiled of their rights and how different identity management in the web is from the real world.

This means that the service provider should ensure to deal with personal identity in one of the chosen mode. However, for security reasons, these modes (e.g., anonymous) should be supersede if an authority (e.g., the police under a mandate from the Justice office) requires so. There is a need for a strong regulation of the digital identity in order to guarantee the fundamental rights of the citizens, but also to allow tracking of users under specific and important (from the perspective of law

[12]See http://www.identitywoman.net/the-identity-spectrum.

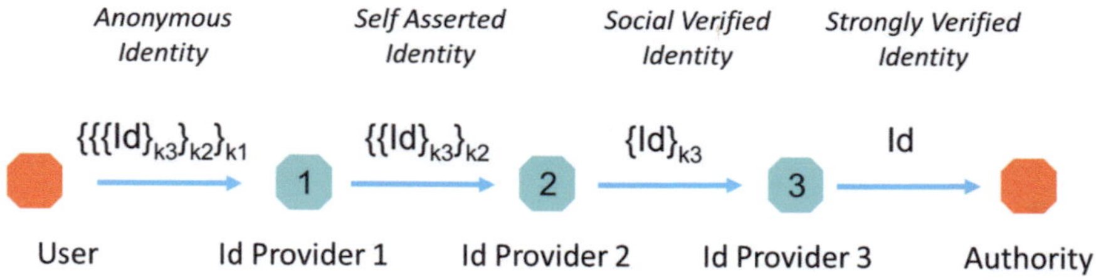

Fig. 3.4 A chain of identity keys

enforcement) conditions. The need for an identity layer has been advocated several times in the Internet community [e.g., (Cameron 2005)], but there is not an agreement in the industry because of the commercial relevance of the issue. In principle the identity should be user-centric, i.e., the user decides session by session the identity mode and associates his identity to a specific identity provider (in certain case it could be the user himself). This layer could be natively integrated in the network in such a way that access to a network will be granted to the individual based on the preferred mode, but with the possibility to strongly identify the user under the request of authorities. The interesting point is that, as happens in most of the social relations in the real world, the user is often his own identity provider by means of assertion about himself. These assertion could be socially validated by friends or people related to the individual. In this way part of the value of the ownership of identity is given back to individuals.

Similar approaches are undertaken by Dark Networks or very specialized P2P networks like i2p (Herrmann and Grothoff 2011) that try to protect and hide the users interactions from control of other parties. Identity management could be dealt within a similar way as information is handled in the i2p network, i.e., through onion routing, and nested encryption by means of a set of private/public keys. Figure 3.4 represents a possible chain of encrypted identifiers/token that can be decrypted by a specific identity provider and passed through to a more secure and low-level Identity provider and enforcer. The last one could be a national authority.

At each step, the identity provider can encrypt only the relevant information at his level; there is no need to disclose more information (that in fact is kept encrypted and is passed to the next Provider, and so forth).

3.8 Cloud Computing and Beyond

The term Cloud Computing is nowadays a common jargon for users and providers. It refers to a computing environment that provides computational services typically in a—client–server fashion. Services can range from on-demand infrastructural capabilities like storage or computing capabilities to applications and services like customer relationship management applications.

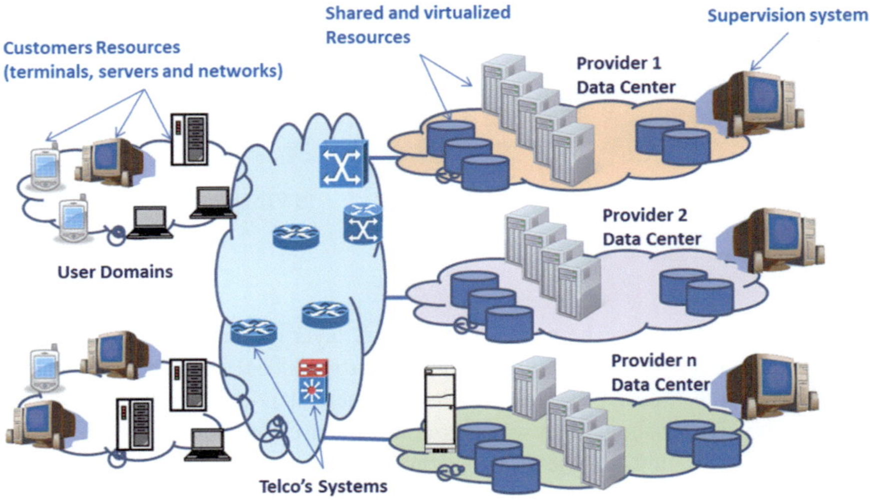

Fig. 3.5 A typical cloud computing configuration

"The National Institute of Standards and Technology (NIST) defines Cloud Computing as a "*pay-per-use model for enabling available, convenient and on-demand network access to a shared pool of configurable computing resources* (e.g., *networks, servers, storage, applications and services*) *that can be rapidly provisioned and released with minimal management effort or service provider interaction*" (Mell and Grance 2011).

Figure 3.5 depicts a typical configuration of a cloud computing infrastructure.

In this representation, user systems can access to "clouds of resources/services/applications" by means of the Internet. Each cloud can provide capabilities on-demand (users buy resources, platforms and services just for the time they need them). Relationships between different clouds can vary according to business relationships among the providers of the infrastructures.

According to NIST (Mell and Grance 2011), a cloud computing system is characterized by a set of essential characteristics, such as:

- On-demand self-service, i.e., the capability offered to a user to directly manage all the needed infrastructure.
- Broad network access, i.e., the ability to access the cloud services by means of common (Internet based) mechanisms independently from the underlying networks (fixed, mobile) and compatibly with the most common devices (PC, mobile phones, tablet and the like).
- Resource pooling, i.e., the providers can dynamically integrate needed resources in order to satisfy customers' needs. Examples of resources are storage, processing, memory, and network bandwidth.

- Rapid elasticity, i.e., the capability to flexibly allocate the needed resources according to availability and customer's demand.
- Measured service, i.e., the providers should make available to customers a precise accounting of resources allocated and used.

The features and capabilities of a cloud system can be summarized into a well renowned model [for instance in Vaquero et al. (2009), Lenk et al. (2009)] that foresees three majors service models

- Software as a service (SaaS), i.e., services and applications are delivered to users by means of a web browser and/or specific client applications.
- Platform as a service (PaaS), i.e., all the typical functionalities of a software platform (e.g., libraries, tools, services) are provided to the users by means of a web browser or a client application.
- Infrastructure as a service (IaaS), i.e., basic capabilities, like processing, storage, and connectivity, are provided to the user that can configure them (e.g., through a web browser of client applications) in order to deploy and execute his/her own services and applications.

From an ecosystem point of view, the NIST definition implies a very simple business model: the pay per use one. It could be implemented by obvious Web companies (like Google and Amazon) or by relevant IT Companies and by Telecom Operators (Telcos).

From a deployment perspective, the NIST definition includes four options

- Private cloud: A full infrastructure (comprising management capabilities) is offered to a single organization.
- Community cloud: The infrastructure is offered and provisioned for exclusive use by a specific community of consumers.
- Public cloud: The cloud infrastructure is offered and provisioned for open use by the general public. This refers mainly to SMEs and residential (but not only) customers.
- Hybrid cloud: The cloud infrastructure is an integration of different cloud infrastructures that remain separated, but are capable of interoperating by means of appropriate technology and business goals.

Figure 3.6 is derived from the taxonomy as defined by NIST.

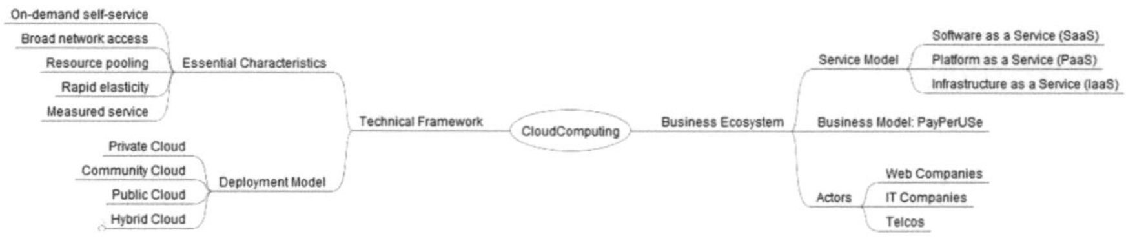

Fig. 3.6 A cloud computing taxonomy that enriches the NIST definition

This chapter provides a wider view at the technological and business level of this cloud computing definition aiming at correctly positioning the Telcos proposition in the market and in the technological scenario.

3.8.1 A Market-Driven Evolution of Cloud Computing

The recent evolution of cloud computing [actually Cloud Computing is a derivative of old and well-known ideas related to utility computing (Vouk 2008)] has been heavily influenced and led from the innovation of web service platforms as provided by big companies like Google, Amazon, and others. These companies have been instrumental in the technological transformation of application servers in very complex (and highly distributed) data centers. They had to put in place highly capable and available data centers able to provide services (e.g., search or selling of goods) to a large audience and with a high variability in demand. Their infrastructure was dimensioned in such ways to be able to provide an answer to each worldwide customer in less than a second. Their infrastructures count for hundreds of thousands of general purpose computing machines (Greenberg et al. 2008).

The opportunistic approach of these giants is based on the fact that they deployed an enormous capacity that is seldom totally used for providing in-house services. There is a lot of spare capacity that they can reuse or can offer to clients. An example is Gmail, the email service offered by Google. In order to index and organize information, the Californian companies had developed over the years a gigantic infrastructure that is capable of storing a large part of the known web. They have spare capacity and they can use it flexibly in order to provide to user large repositories for collecting mails. The variance between the deployed capabilities and the real usage of them is the key for providing cloud computing services. Figure 3.7 depicts two typical situations, the first one in which the total capacity is always greater than the demand for resources (in this case the business goal is to sell the spare capacity, i.e., shown as flexibility in Case 1); the second one depicts a situation in which sometimes all the resources are over allocated, a sort of negative flexibility that can hamper the functioning of the system (in this case a better allocation strategy is to be implemented, e.g., able to optimize service language agreements (SLAs) and to reduce penalties).

3.8.2 Some Technology Trends in Cloud Computing

The technological panorama of cloud computing is vast, in this section a few aspects of its evolution will be taken into consideration. They are relevant for understanding how the current market propositions are put forward by major actors and their implications from a service ecosystem point of view.

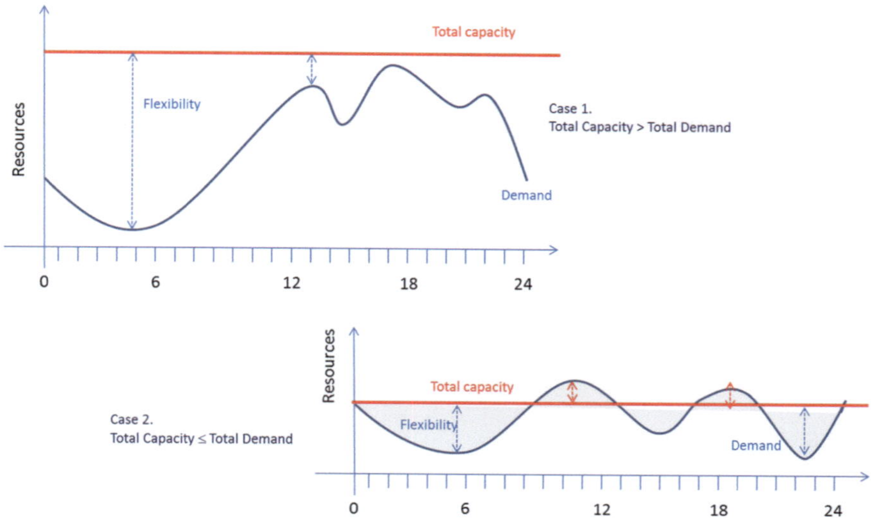

Fig. 3.7 Flexibility in capacity allocation in data centers

Emphasis on Data Management

The current offering of services and applications in cloud computing is derived largely by the capability of some companies in dealing with huge datasets. Two of the major actors in this field, namely Google and Amazon, were pioneers of new ways for dealing with large datasets and the use of advanced techniques. Google has been using the MapReduce approach (Dean and Ghemawat 2008) in order to implement an indexing mechanism able to perform on a large infrastructure of general-purpose machines. The MapReduce[13] method consists of a Map() procedure that filters and sorts a sequence of data and a Reduce() procedure that executes a combination of available results of Map operations. There are two steps in this method

"Map" step: The master node takes the input, splits it into smaller subproblems, and allocates them to worker nodes. The worker node executes the smaller problem, and returns the answer to the master node.
"Reduce" step: The master node then gathers the answers to all the subproblems and combines them to form the result.

In this way, large set of data are reduced into smaller chunks and each chunk is dealt with in parallel by a worker. Intermediate results are sorted out and combined by reduce processes in an ordered sequence (as depicted in Fig. 3.8).

The Google approach was inspirational and this has led to the well-known approach of the Hadoop open source platform (White 2010). MapReduce and

[13]See for instance http://en.wikipedia.org/wiki/MapReduce. Last accessed May 30th 2013.

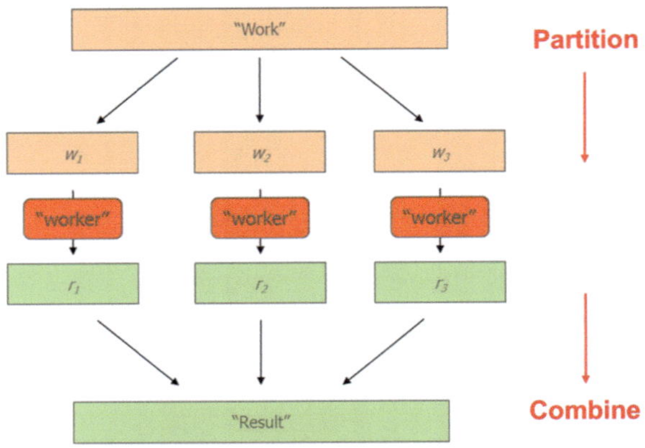

Fig. 3.8 A MapReduce example

Hadoop mechanisms are often offered as a service in cloud computing platforms [e.g., Amazon Elastic MapReduce Service (Gunarathne et al. 2010)].

Another example of data processing innovation is the Amazon platform. It is intended to support the large variety of services offered by Amazon (see Fig. 3.9).

Particularly interesting in the realm of data manipulation is the Amazon solution for the simple storage service (S3). Dynamo (DeCandia et al. 2007) is a highly available, proprietary-distributed storage system. Its goal is to provide database services in a highly distributed environment. In order to reach the goal, it is based on a key-value approach and it uses distributed hash tables (DHTs) for pointing to

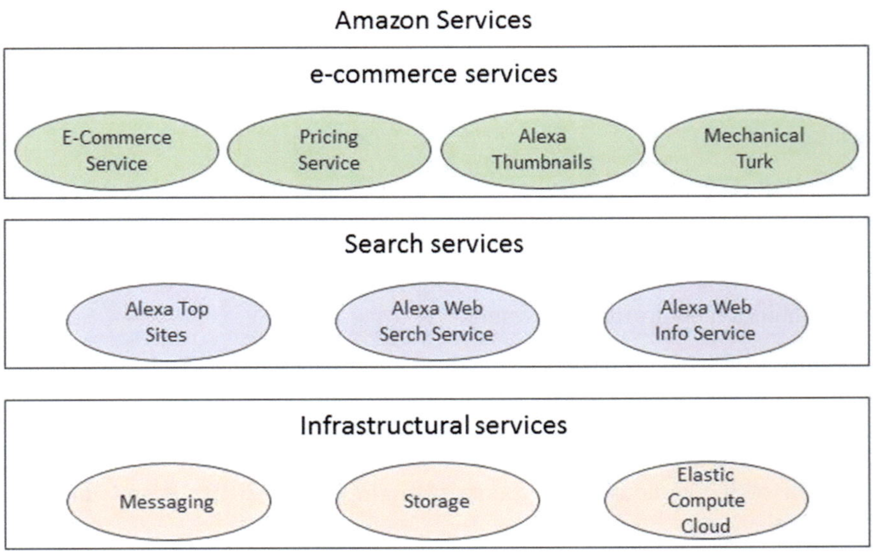

Fig. 3.9 The Amazon cloud infrastructure

data. Functions similar to Dynamo are offered through SimpleDB web service (Sciore 2007) by Amazon (Murty 2009), which also offers elastic basic storage services through S3 (Brantner et al. 2008). Dynamo is one example of a consistent trend in database evolution named NoSQL, it is not following a traditional relational database management system approach, and instead it supports the high distribution and partition of huge datasets by means of a distributed, fault-tolerant architecture. Another example of this trend is the Facebook's internal platform Cassandra and the already cited MapReduce and Hadoop systems. Actually the design choices for dynamo are: scalability in order to add new systems to the network minimizing their impact; symmetry, i.e., each node has no special roles, in fact all features are in all nodes; decentralization, the dynamo design do not foresee any master node(s); high availability, data are replicated in the network nodes and they must be always available; speed, the system should provide access to distributed data very quickly and consistently with user requirements. These design guidelines have privileged partition and availability of data over the consistency of data (or better: dynamo adopted a "weak consistency" model according to Brewer's CAP theorem (Brewer 2012).[14] In other terms, data will be distributed and always available even if replicas of data could be in different (and inconsistent) states. Another interesting feature is that data are always writable because conflicts are dealt with during "reads."

Another relevant example of availability and timeliness in providing information is given by the functions implemented by the Twitter architecture that is able to provide in real-time information about the activities and the information that users perform or share within the system. In this case, the problem is not only indexing the information in order to allow a fast retrieval to users, but also to tag the information and make it available in the shorter time possible to user that are checking specific hashtags. In fact the Twitter engine is based on a PubSub model (Fontoura et al. 2013) and each single event to be published is indexed and delivered to interested users. The Twitter architecture integrates a message queue engine, Kestrel, a Hadoop base content store with a NoSQL metadata store based on the Cassandra solution (Lakshman and Malik 2010). Also in this service, consistency of data is a minor requirement compared to quasi real/time availability and partition.

Generally speaking, an approach that favors the availability instead of consistency can deal with huge data and can provide very fast response time by discounting the needs of consistency of all the data replicas. However, the consistency requirement could be a major need (e.g., in financial and transactional-related applications). In these cases, a network able to provide high availability features

[14]From Wikipedia: "Brewer's theorem, states that it is impossible for a distributed computer system to simultaneously provide all three of the following guarantees: (a) Consistency (all nodes see the same data at the same time); (b) Availability (a guarantee that every request receives a response about whether it was successful or failed); (c) Partition tolerance (the system continues to operate despite arbitrary message loss or failure of part of the system). According to the theorem, a distributed system cannot satisfy all three of these guarantees at the same time.

and to keep delays within specific intervals could be needed. Depending on services, consistency could be pursued in high-distributed systems by relaxing requirements either on high availability or on partition of data (centralization). In this case, appropriate control paradigms should be chosen: dynamo and big table have chosen P2P, Twitter has chosen PubSub; consistent networking could be supported by the Network Intelligence paradigm.

Virtualization

Virtualization is widely used in cloud computing solutions. Actually the progress of the virtualization techniques is not penalizing too much from the performance point of view. The concept of hypervisor and related technologies have matured so much that now different options are possible and each of them is not penalizing too much the overall performance of the hosting machines. Typical configurations are depicted in Fig. 3.10.

Virtualization has been applied mainly to processing and storage changing the face of utility computing and taking progressively advantage of multicore systems (Kumar et al. 2007).

OpenNebula (Sempolinski and Thain 2010) is an interesting open source solution because it provides a uniform and homogeneous view of virtual resources, abstracting away from different virtualization technologies by means of drivers (new drivers can be created to add support to new virtualization technologies). It uses a scheduler, which can be easily tailored or changed, to take virtual machines (VM) placement decisions (e.g., to balance the workload or to consolidate servers). In addition it is able to support the virtual machines migration and this makes openebula a good solution for experimentation in a highly dynamic environment like those inspired to P2P or those posed at the edge of the network and supported by edge nodes.

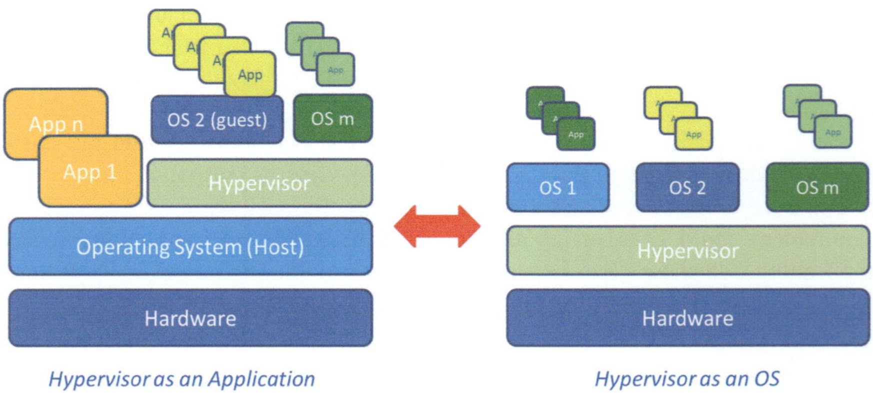

Fig. 3.10 Examples of hypervisor applicability

Focalization on Perimeterized Solutions

Cloud computing comes also with a number of drawbacks, for example clouds are designed to interact within a homogeneous environment. Usually providers prefer to impose a close and proprietary environment instead of looking for interoperability with other systems. Providers of cloud solutions have generally adopted a walled garden approach, i.e., a cloud is internally homogeneous in terms of approach, machine virtualization, interfaces, and it is not interoperable with other clouds. This implies that the user that needs services from different clouds has to cope with interoperability issues within the customer domain (e.g., interoperability between resources in Cloud Domain 1 and in Cloud Domain N has to be sorted out within the customer domain). Figure 3.11 depicts one of such situations in which different clouds are homogeneous (right) but they cannot interwork directly (left) and it is a task of the user to integrate the functions provided by the two different cloud services.

Actually the interoperability is a major difference between cloud and grid computing (Dillon et al. 2010), the latter is more complicated in terms of interfaces and mechanisms but it can support interworking of heterogeneous systems. The concept behind grid computing is the possibility to put together heterogeneous resources provided by multiple providers in order to integrate them into a virtual organization. Resources are negotiated for and are chosen according to specific needs of the applications. Resources can be used as single elements or they can be aggregated and used as collections/groups. In addition, grid computing is standardizing a set of programming interfaces in order to allow the development of customized applications and services fulfilling particular needs and requiring specific arrangements of resources. Obviously this architectural characteristic comes with a price: programming at collective or single resource level implies more complexity and a clear knowledge of how to compose and organize resources.

On the other side, in cloud computing, the interoperability at customer level can be alleviated by means of virtualization, in fact if all the involved domains are providing virtualized resources of the same type, the customer applications can

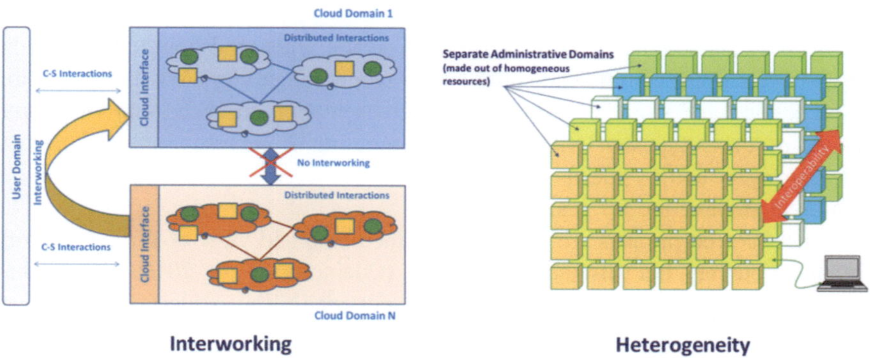

Fig. 3.11 Cloud computing: current interoperability (*left*) and heterogeneity (*right*)

have a homogenous view on available resources independently from the specific cloud. In any case, the creation of walled gardens is a market strategy of cloud providers in order to segment and perimeterize the cloud offering.

Lack of solutions considering network issues

Another major issue of cloud computing is the lack of any references to the underlying networks. Networks are abstracted and networking resources are not made visible to users and applications by means of interfaces. This diminishes the flexibility of the cloud because the integration of processing, storage, and communications allows building platform that are more capable to adapt to the dynamic needs of the applications imposed by the execution context. The assumption is that connectivity is granted and it will be provided according to the best effort arrangement. It is also assumed that the capacity of the supporting networks is sufficient for services and applications to deliver the expected behavior. Actually this could be a big issue as pointed out in Deutsch (1995) and Rotem-Gal-Oz (2006). The assumption that the network per se will always be available providing the expected services is wrong and dangerous and it could lead to disruptive effects on services. Many cloud computing solutions are designed in such a way to cope with dynamic behavior of networks and they try to trade off the unreliable behavior of the network by increasing processing and storage capabilities at the edges (in the user domain and in the cloud). As an example of this trend, the Amazon Silk browser dynamically splits computation between the servers and the end device (in this case a tablet) in order to optimize the resources and the processing load between components. This is done to mediate the adverse cases of a malfunctioning network status. Actually some programming languages like Ambient Talk (Dedecker et al. 2006) have been designed in order to cope with network fallacies. It is based on the possibility of communication processes to keep working while hiding to the programmers the need to check the network connectivity status.

A major difference between cloud and grid computing is the view on resources: the grid has defined and uses interfaces that allow to manage connectivity according to the needs of the distributed applications. If an application needs to transfer a large amount of data between different computing nodes (e.g., specialized in the analysis of medical images), this requirement can be better fulfilled if the application can control the connectivity between nodes and decide when and how to transfer the files. While current commercial solutions for cloud computing are mainly focusing on processing and storage (giving simple representation and access to virtualized images of these types of entities), grid computing is representing resources at different layers and by means of well-defined interfaces. Figure 3.12 represents a high-level architecture for grid computing.

Some research projects are investigating the combined view of IT and network virtual resources. For instance, IRMOS[15] (Cucinotta et al. 2010) project has introduced the concept of Virtual Service Network (VSN), which consists in an

[15]See http://www.irmosproject.eu/.

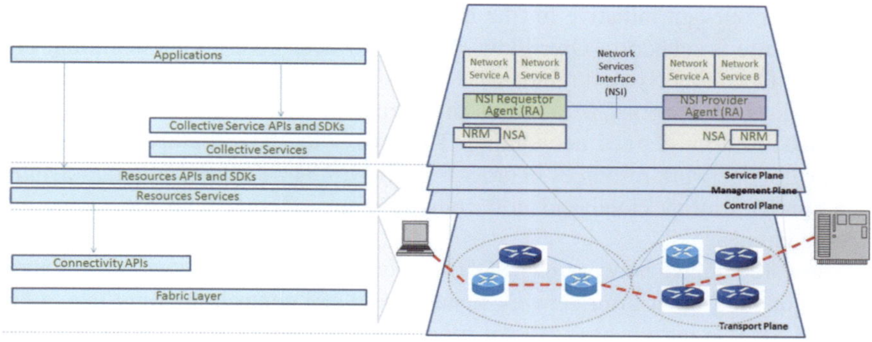

Fig. 3.12 Grid computing layered model and the network representation

aggregation of VMs, virtual links, and virtual storage nodes. The VSN description, an ontology-based graph model (OWL), specifies hardware and QoS requirements for each of the virtual resources integrating the network. In the IRMOS platform, a VSN is created for each application as part of an automatic SLA negotiation process. During such process, an ISONI provider (i.e., provider of virtual resources) able to fulfill the hardware and QoS requirements of the application is dynamically discovered and selected (all virtual resources will be provided by the same provider). Resources are then reserved and integrated into the VSN where the application will run. At run time, the initially signed SLA can be renegotiated, but only the amount of resources assigned to the VSN can be modified (not the type of resources, and not their provider). SLA renegotiation and the subsequent VSN modification can be triggered because one of the following: the application end-user changed the QoS requirements of the application, the initial amount of reserved resources was not enough to satisfy the signed SLA, or as a response to some scalability rules specified for the application.

Applications at the upper layer have different interfaces that can be used to control and manage resources (or group of resources termed as collective) at different layers and with different granularity. In this way an application can negotiate and allocate a collective resource (an aggregate of functionalities provided by a group of resources that can be controlled and managed as a single entity), or it can access to resources and to connectivity for linking them together. In the specification (Roberts 2010) of the Open Grid Forum, a model for a network of network is given in order to present to applications and services an interface for requesting and controlling the composition of connectivity among different networks. A simpler specification (Metsch and Edmonds 2011) is provided within the framework of OGF by the Open Cloud Computing Interface, OCCI initiative. It provides an object model for describing how resources can be connected by means of links and network interfaces. The goal is to allow applications and Cloud Infrastructures to cope with the complexity of supporting networks and to orchestrate the needed resources on a dynamic basis.

Another relevant initiative for the integration of network virtualization in cloud infrastructure is related to OpenFlow (McKeown 2008) and to the definition of a Network Operating System (Gude et al. 2008) and its applicability to data centers (Tavakoli et al. 2009). OpenFlow is the parent of new initiatives related to the so-called software-defined networking. The goal is to allow the opening up interfaces within network resources in order to allow virtualization and programmability in a similar way as cloud and grid computing are offering with processing and storage entities.

Software-defined networking (SDN) is about virtualizing network equipment and decoupling them from network management and control; not only this, a key facet of SDN is introducing API for programming network services. In principle, this could mean morphing routers into commodity (low cost) programmable boxes controlled and programmed (through API) by an outside source. This research track may have a deep impact on cloud computing. For example, in Armbrust et al. (2010), it is mentioned how two-thirds of the cost of WAN bandwidth is the cost of the high-end routers, whereas only one-third is the fiber cost. So, simpler network nodes (e.g., routers) built from commodity components (as SDN is planning to have) deployed in WAN, may provide costs dropping more quickly than they have had historically, enabling new paradigms of interactions Cloud—Network.

In this direction, OpenStack (Pepple 2011) is an open source cloud project and community with broad commercial and developer support. OpenStack is currently developing two interrelated technologies: OpenStack Compute and OpenStack Object Storage. OpenStack Compute is the internal fabric of the cloud creating and managing large groups of virtual private servers and OpenStack Object Storage is software for creating redundant, scalable object storage using clusters of commodity servers to store terabytes or even petabytes of data. Interestingly, OpenStack has a network connectivity project named Quantum.[16] Quantum looks to provide "network connectivity as a service" between interface devices managed by other OpenStack services. Quantum itself does not talk to nodes directly: it is an application-level abstraction of networking. It requires additional software (in the form of a plug-in) and it can talk to SDN via an API.

3.9 Terminals and Devices Evolution

The basic key technologies (see Table 3.1) having an impact on the evolution of terminals are rapidly progressing and in the future years, they will have meaningful effects on terminals. For instance

[16]Information available at https://wiki.openstack.org/wiki/Quantum. Last accessed May 29th 2013.

- New Radio Interfaces (beyond LTE advanced) and multicore technologies will bring the capability to have more bandwidth, in fact each core could process flows at a specific frequency, and the aggregation of different point to point connections at specific frequencies will exceed the current limitations;
- Each terminal will be capable of handling different technologies for exploiting local communications;
- Always best connected services will be dictated by the terminals that will use available spectrum (and cognitive radio capabilities) to get the best connectivity possible in the operation context;
- New Interfaces like touch, voice, gestures, and bio-neural will allow the creation of natural and easy to use interfaces. Some interfaces could be portable from terminals to terminals becoming a distinctive market offering;
- M2M (machine-to-machine) and IoT (Internet of Things) capabilities will be such that each terminal will be part of crowdsensing communities, i.e., aggregation of terminals that gather information on the context and the environment and return it to a community that will use it to provide services or simply monitor a part of a city;
- Displays (e.g., HD, 3D, Flexible, pico projectors) will support a high quality almost indistinguishable from the reality and this will enable new services. In addition the terminal itself will be able to project in the environment video and images;
- Disappearing of SIMs; Identity will be based on different mechanisms (bio-Id) that will allow for higher levels of security, profiling, services because the terminal will be capable to recognize the actual user and to setup the personal environment in cooperation with cloudified information and applications;
- Storage capabilities in terminals will continue to progress and it will be abundant also in low-level terminals. The combination of large storage capabilities, availability of connectivity and function provided in the cloud, and increased processing power as well as multimedia features will make possible to record almost any moment of life of users;
- Smart materials and nanotechnologies will allow the creation of flexible, adaptable and wearable terminals that will act also as sensors for users;
- Batteries limitations will remain and likely will be the major barrier to further developments.

Terminals will be more and more “personal”. They will be worn all the time, their components will be taken apart (disassembled), and made extremely portable (some components could be deployed in or use the human body or clothes). Almost any aspect of personal life will be measured and stored. The positioning of Internet companies in the control of terminals/services will become even more stringent and there will be the need to have regulation that safeguard privacy and establish fair rules for the usage of data captured by terminals. The current battle for supremacy in the mobile operating system sector will have taken new forms, but it will be fundamental also in the future. Likely, application portability and the breaking of closed walled garden will become a necessity for users, but also for the industry.

There will be a movement for the standardization of terminal capability that could lead to a situation similar to the current web, i.e., a few stable protocols and mechanisms that enable users to take advantage of the web. In this context there will be an increasing synergy between the terminals and the cloud in order to make readily available to the users all their data and environments. Some browsers (like Amazon's Silk and Apple iCloud) are now already exploiting synergies with cloud applications, this trend will consolidate even more in the future (Bahl et al. 2012).

Users will strive to access in the same way the same service from several different terminals and will look for the same user experience. Virtualization of terminals and applications within the cloud will be a possible trend that will enable the users to break the walled garden approach taken so far by mobile terminal vendors.

Terminal will also be the means to provide augmented and virtual reality applications that will make possible the creation of new compelling and immersive services. These services should be greatly customized in order to present to users the desired information.

Terminals will also be a major means to support new forms of payment, from prepaid card systems up to newer ones. Virtual and digital money will be a major issue. So far all the money transaction on the Internet are tracked, there is not the possibility to spend money in an anonymous way for buying goods like it happens in the real world. Possibly this will change if digital currencies like Bitcoin (Barber et al. 2012) will have success. In any case, mobile terminals will be the new wallets of the future.

The combination of all these features and challenges will make the terminals even more central, if possible, in future communications ecosystems. Having a grip on terminals will mean essentially to have a chance to own the users. On the other side, the progress of technologies and the awareness of users and regulators should also help in limiting unfair usage of portable devices. In any case, users should acquire much more awareness of their possibility to control this fundamental element of the ICT ecosystem.

3.10 Findings

As seen in Sect. 3.2, the evolution of basic technologies will continue at the same pace of today. In a ten years timeframe, the processing power will double every 18 months leading to an increase of more than 50 times while the power consumption will be cut down. Solutions like RasperryPi will not be an exception and be widely deployed. Storage will follow a similar path. Screen technologies will support a "natural" interaction with humans. Pervasiveness of computing will be supported by the wide availability of low cost connectivity granted by improved local capabilities and the availability to manage frequencies at the local level in freer modes. These developments give a clear idea of the possibilities offered by tomorrow technologies well beyond the current status. Data centers will become commodities and distribution of functionalities over different administrative domains will become a stringent requirement. Virtualization will allow the creation of adaptive and virtualized computing environments. Their possibilities will be further considered and described in Sect. 5.5.

Data will play a fundamental role: owning data and being able to infer information from them will be a distinctive advantage of some Actors. This advantage should be mitigated by appropriated and fair policies for guaranteeing to users (to be interpreted as citizens and customers) some levels of privacy and ownership on their data.

With respect to data center and cloud computing technologies, this chapter has shown that WebCos have a technology advantage over many other actors. This advantage is built on the C-S approach as an interaction paradigm (simplicity and efficiency in getting the service functionalities). It will be difficult to close the gap with them for any company willing to have a predominant role in the "cloud". These companies are advancing the technologies behind the server front end toward highly distributed systems preparing for "switching" to even further level of distribution of functions. Following them on this terrain is extremely risky, time consuming, and investment heavy: Actors like Google and Amazon have made their technical infrastructure an asset that is daily increased by adding new functions and new solutions. In addition they have a total control on the systems having built it piece by piece. Telcos have not this capability and this fact has consequences as the reduced ability to program the "network platform", the lack of innovation, and the lack of competences in the software development.

The evolution of terminals is driven by some major factors: the progress of basic technologies, the disposition of users to invest into terminals, the need of terminals in order to access the services. Terminals are now seen by Webcos as a means to control the client side of the C-S model. Entering in this area allows them to have a full control on the value chain of the services leaving to Telcos only the commoditized part of the connectivity.

In this evolution, the user is constrained to a passive role: consumer of services, consumer of terminals, and consumer of data. Users can contribute to services in terms of profiles and user generated content. However the increasing capabilities of terminals could be an element that will give some freedom to users: in fact mobile terminals can be used to directly communicate (Direct to Direct communication can be seen as the new frontier of some communications services that further disrupt the Telcos market); they can be used to aggregate into local clouds capable of elaborating local data and providing local services to the participants; terminals can collect and process large quantities of personal data preserving privacy of users. Terminals capabilities will be fundamental in order to access and support functionalities of highly adaptive and dynamic smart environments.

In a linear projection of technology evolution, the power of current solutions will allow Webcos to keep an edge over the other competitors: services could be considered as a natural basin of these companies. This has the consequence that the role of Telcos will remain essentially the same also for the future: traditional Telcos or connectivity providers. The technologies and the solutions as well as the business motivations behind these scenarios will be further considered in Chap. 6.

However the growing capabilities of terminals, the increasing awareness of some individuals for a different approach to privacy and management of personal data could be interpreted as two possible inflection points capable of determining new paths in the evolution of the ICT industry. These two capabilities are important because they may enable new scenarios for the service provision strongly dominated by edge computing capabilities. There is a broad possibility in this area and exploiting the edge resources could lead to a disruption in the traditional C-S approach both as a technical and business paradigm. Telcos could try to take a chance to operate in this sector in order to regain technical disruption and to operate closer to users.

Bibliography

Armbrust M, Fox A, Griffith R, Joseph AD, Katz R, Konwinski A et al (2010) A view of cloud computing. Commun ACM 53(4):50–58

Auer S, Bizer C, Kobilarov G, Lehmann J, Cyganiak R, Ives Z (2007) Dbpedia: A nucleus for a web of open data. In: The semantic web. Springer, pp 722–735

Bahl P, Han RY, Li LE, Satyanarayanan M (2012) Advancing the state of mobile cloud computing. In: Proceedings of the third ACM workshop on Mobile cloud computing and services. ACM, Low Wood Bay, Lake District, United Kingdom, pp 21–28

Barber S, Boyen X, Shi E, Uzun E (2012) Bitter to better—how to make bitcoin a better currency. In: Keromytis AD (ed) 16th conference on financial cryptography and data security. Springer, Kralendijk, pp 399–414

Baroncelli F, Martini B, Valcarenghi L, Castoldi P (2005) A service oriented network architecture suitable for global grid computing. In: Conference on optical network design and modeling. IEEE - IP TC6, Milan, Italy, pp 283–293

Barroso LA, Dean J, Holzle U (2003) Web search for a planet: The Google cluster architecture. IEEE Micro 23(2):22–28

Bifet A, Frank E (2010) Sentiment knowledge discovery in twitter streaming data. In: Pfahringer B, Holmes G, Hoffmann A (eds) 13th international conference on discovery science. Springer, Canberra, pp 1–15

Brantner M, Florescu D, Graf D, Kossmann D, Kraska T (2008) Building a database on S3. In: Proceedings of the 2008 ACM SIGMOD international conference on Management of data. ACM, pp 251–264

Brewer E (2000) Towards robust distributed systems. In: Proceedings of the annual ACM symposium on principles of distributed computing, vol 19. ACM, Portland, p 7

Brewer E (2012) Pushing the CAP: strategies for consistency and availability. Computer 45(2): 23–29

Burrows M (2006) The Chubby lock service for loosely-coupled distributed systems. In: Proceedings of the 7th symposium on Operating systems design and implementation. USENIX Association, Berkeley, pp 335–350

Buyya R, Ranjan R, Calheiros RN (2010) Intercloud: Utility-oriented federation of cloud computing environments for scaling of application services. In: ICA3PP'10 Proceedings of the 10th international conference on Algorithms and Architectures for Parallel Processing. Springer, pp 13–31

Cameron K (2005) The laws of identity. Microsoft

Cattell R (2011) Scalable SQL and NoSQL data stores. ACM SIGMOD Rec 39(4):12–27

Chang F, Dean J, Ghemawat S, Hsieh WC, Wallach DA, Burrows M et al (2008) Bigtable: a distributed storage system for structured data. ACM Trans Comput Syst (TOCS) 26(2), article n 4

Clark DD, Wroclawski J, Sollins KR, Braden R (2005) Tussle in cyberspace: defining tomorrow's internet. IEEE/ACM Trans Networking (TON) 13(3):462–475

Cucinotta T, Checconi F, Kousiouris G, Kyriazis D, Varvarigou T, Mazzetti A et al (2010) Virtualised e-learning with real-time guarantees on the irmos platform. In: IEEE international conference on Service-Oriented Computing and Applications (SOCA). IEEE, Perth, pp 1–8

Dean J, Barroso LA (2013) The tail at scale. (ACM, ed.). Commun ACM 56:74–80

Dean J, Ghemawat S (2008) MapReduce: simplified data processing on large clusters. Commun ACM 51(1):107–113

Dean J, Ghemawat S (2010) MapReduce: a flexible data processing tool. Commun ACM 53 (1):72–77

DeCandia G, Hastorun D, Jampani M, Kakulapati G, Lakshman A, Pilchin A et al (2007) Dynamo: amazon's highly available key-value store. In: ACM Symposium on Operating Systems Principles: Proceedings of twenty-first ACM SIGOPS symposium on Operating systems principles, vol 14. ACM, pp 205–220

Dedecker J, Van Cutsem T, Mostinckx S, D'Hondt T, De Meuter W (2006) Ambient-oriented programming in ambienttalk. In: ECOOP 2006–Object-Oriented Programming. Springer, Nantes, pp 230–254

Deutsch P (1995) Fallacies of Distributed Computing. Wikipedia

Di Costanzo A, De Assuncao MD, Buyya R (2009) Harnessing cloud technologies for a virtualized distributed computing infrastructure. IEEE Internet Comput 13(5):24–33

Dillon T, Wu C, Chang E (2010) Cloud computing: issues and challenges. In: 24th IEEE international conference on Advanced Information Networking and Applications (AINA). IEEE, Perth, pp 27–33

Fontoura A, Maxim S, Marcus G, Josifovski V (2013) Top-k publish-subscribe for social annotation of news. In: Proceedings of the VLDB Endowment. VLDB, p. to appear

Foster I, Zhao Y, Raicu I, Lu S (2008) Cloud computing and grid computing 360-degree compared. In: Grid computing environments workshop. IEEE, Austin, pp 1–10

Ghemawat S, Gobioff H, Leung S-T (2003) The Google file system. ACM SIGOPS Operating Syst Rev 37(5):29–43

González MC, Hidalgo C, Barabasi A-L (2008) Understanding individual human mobility patterns. Nature 453(7196):779–782

Greenberg A, Hamilton J, Maltz DA, Patel P (2008) The cost of a cloud: research problems in data center networks. ACM SIGCOMM Comput Commun Rev 39(1):68–73

Gude N, Koponen T, Pettit J, Pfaff B, Casado M, McKeown N et al (2008) NOX: towards an operating system for networks. ACM SIGCOMM Comput Commun Rev 38(3):105–110

Gunarathne T, Wu T-L, Qiu J, Fox G (2010) MapReduce in the clouds for science. In: IEEE second international conference on cloud computing technology and science (CloudCom). IEEE, Indianapolis, pp 565–572

Herrmann M, Grothoff C (2011) Privacy-implications of performance-based peer selection by onion-routers: a real-world case study using I2P. In: PETS'11 11th international conference on Privacy enhancing technologies. Springer, Waterloo, pp 155–174

Hey AJ, Tansley S, Tolle KM et al (2009) The fourth paradigm: data-intensive scientific discovery. Microsoft Research, Redmont

Hey AJ, Tansley S, Tolle KM et al (2011) The fourth paradigm: data-intensive scientific discovery. Proc IEEE 99(8):1334–1337

Hsu PP, Mulay V (2011) Open Compute Project–Server and Data Center Design. Silicon Valley Leadership Group SVLG

Izal M, Urvoy-Keller G, Biersack EW, Felber PA, Al Hamra A, Garces-Erice L (2004) Dissecting bittorrent: five months in a torrent's lifetime. In: Barakat CA (ed) Passive and active network measurement—lecture notes in computer science, vol 3015. Springer, Berlin, pp 1–11

Koomey JG (2010) Outperforming Moore's law. IEEE Spectr 47(3):68

Krishnamurthy B, Gill P, Arlitt M (2008) A few chirps about twitter. In: Proceedings of the first workshop on online social networks. ACM, Seattle, pp. 19–24

Kumar S, Raj H, Schwan K, Ganev I (2007) Re-architecting VMMs for multicore systems: the sidecore approach. In: Workshop on interaction between opearting systems and computer architecture (WIOSCA). University of Florida, San Diego, p. paper 3
Lakshman A, Malik P (2010) Cassandra: a decentralized structured storage system. ACM SIGOPS Operating Syst Rev 44(2):35–40
Landau S, Moore T (2011) Economic tussles in federated identity management. In: Proceedings of the tenth workshop on the economics of information security (WEIS). George Mason University, USA
Lantz B, Heller B, McKeown N (2010) A network in a laptop: rapid prototyping for software-defined networks. In: Proceedings of the 9th ACM SIGCOMM workshop on hot topics in networks. ACM, Monterey, p. article n 19
Laoutaris N, Rodriguez P, Massoulie L (2008) ECHOS: edge capacity hosting overlays of nano data centers. ACM SIGCOMM Comput Commun Rev 38(1):51–54
Leavitt N (2010) Will NoSQL databases live up to their promise? IEEE Comput 43(2):12–14
Lenk A, Klems M, Nimis J, Tai S, Sandholm T (2009) What's inside the Cloud? An architectural map of the Cloud landscape. In: Proceedings of the 2009 ICSE workshop on software engineering challenges of cloud computing. IEEE, Vancouver, pp 23–31
Malewicz G, Austern MH, Bik AJ, Dehnert JC, Horn I, Leiser N et al (2010). Pregel: a system for large-scale graph processing. In: Proceedings of the 2010 ACM SIGMOD international conference on management of data. ACM, Indianapolis, pp 135–146
Marshall P, Keahey K, Freeman T (2010) Elastic site: Using clouds to elastically extend site resources. In: CCGRID '10 proceedings of the 2010 10th IEEE/ACM international conference on cluster, cloud and grid computing. IEEE Computer Society, Melbourne, pp 43–52
Mathioudakis M, Koudas N (2010) Twittermonitor: trend detection over the twitter stream. In: Proceedings of the 2010 international conference on management of data. ACM, Indianapolis, pp 1155–1158
McKeown NT (2008) OpenFlow: enabling innovation in campus networks. ACM SIGCOMM Comput Commun Rev 38(2):69–74
Mell P, Grance T (2011) The NIST definition of cloud computing (draft). NIST, Washington
Metsch T, Edmonds A (2011) Open cloud computing interface—infrastructure. Open Grid Forum, Lemont
Miller KC, Brandwine JE, Doane AJ (2011) Using Virtual Networking Devices to manage routing communications between connected computer networks. United States Patent Office
Minerva R, Crespi N (2011) Unleashing the disruptive potential of user-controlled identity management. In: Telecom World (ITU WT), 2011 Technical Symposium at ITU. IEEE, Geneva, pp 1–6
Minerva R, Demaria T (2006) There is a Broker in the Net… Its name is Google. In: Proceeding of ICIN conference. ADERA—ICIN, Bordeaux, pp 1–6
Minerva R, Manzalini A, Moiso C (2011) Towards an expressive, adaptive and resource aware Network Platform. In: Prasad A, Buford J, Gurbani V (eds) Advances in next generation services and service architectures. River Publisher, pp 43–63
Minerva R, Moiso C, Manzalini A, Crespi N (2013) Virtualizing platforms. In: Bertin E, Crespi N, Magedanz T (eds) Evolution of telecommunication services, vol 7768. Springer, Berlin, pp 203–226
Moiso C, Minerva R (2012) Towards a user-centric personal data ecosystem the role of the bank of individuals' data. In: 16th international conference on intelligence in next generation networks (ICIN). IEEE, Berlin, pp 202–209
Murty J (2009) Programming amazon web services: S3, EC2, SQS, FPS, and SimpleDB. O'Reilly Media, Sebastopol
NFV (2012) Network functions virtualisation. ETSI, Sophia Antipolis
Nielsen J (1998) Nielsen's law of internet bandwidth. Useit
Ohtsuki H, Hauert C, Lieberman E, Nowak AM (2006) A simple rule for the evolution of cooperation on graphs and social networks. Nature 441(7092):502–505

Open Networking Foundation (2012) OpenFlow-Enabled cloud backbone networks create global provider data centers. Open Networking Forum, Palo Alto
Pepple K (2011) Deploying OpenStack. O'Reilly Media, Sebastopol
Piatetsky-Shapiro G (2007) Data mining and knowledge discovery 1996 to 2005: overcoming the hype and moving from "university" to "business" and "analytics". Data Min Knowl Disc 15 (1):99–105
Pike R, Dorward S, Griesemer R, Quinlan S (2005) Interpreting the data: parallel analysis with Sawzall. Scientific Programming 13(4):277–298
Platform G, Google platform. Wikipedia. Last accessed May 2013
Roberts G, Kudoh T (2010) Network services framework v1.0. Open Grid Forum, Lemont
Rotem-Gal-Oz A (2006) Fallacies of distributed computing explained. http://www.rgoarchitects.com/Files/fallacies.pdf
Schaller RR (1997) Moore's law: past, present and future. IEEE Spectr 34(6):52–59
Sciore E (2007) SimpleDB: a simple java-based multiuser syst for teaching database internals. ACM SIGCSE Bull 39(1):561–565
Sempolinski P, Thain D (2010) A comparison and critique of eucalyptus, opennebula and nimbus. In: IEEE second international conference on cloud computing technology and science (CloudCom). IEEE, Indianapolis, pp 417–426
Stonebraker M (2010) SQL databases v. NoSQL databases. Commun ACM 53(4):10–11
Stonebraker M, Cetintemel U, Zdonik S (2005) The 8 requirements of real-time stream processing. ACM SIGMOD Rec 34(4):42–47
Tavakoli A, Casado M, Koponen T, Shenker S (2009) Applying NOX to the datacenter. In: HotNets. ACM, New York
Vahdat A (2012) Symbiosis in scale out networking and data management. In: Proceedings of the 2012 international conference on management of data. ACM, Scottsdale, pp 579–580
Valancius V, Laoutaris N, Massoulie L, Diot C, Rodriguez P (2009) Greening the internet with nano data centers. In: Proceedings of the 5th international conference on Emerging networking experiments and technologies. ACM, Rome, pp 37–48
Vaquero LM, Rodero-Merino L, Caceres J, Lindner M (2009) A break in the clouds: towards a cloud definition. ACM SIGCOMM Comput Commun Rev 39(1):50–55
Vouk MA (2008) Cloud computing—issues, research and implementations. In: 30th International conference on information technology interfaces, 2008. ITI 2008. IEEE, Dubrovnik, pp 31–40
Walter C (2005) Kryder's law. (S. American, ed.). Sci Am 293(2):32–33
White T (2010) Hadoop: the definitive guide. O'Reilly Media Inc, Sebastopol
World Economic Forum (2011) Personal data: the emergence of a new asset class. World Economic Forum, Cologny, Geneva
Zaslavsky A, Perera C, Georgakopoulos D (2013) Sensing as a service and big data. arXiv preprint

Chapter 4
The New Network

The future network must gear up to cope with massive data growth, but also provide what is not available on the Internet today—security, reliability, mobility, and predictability. Current Internet structures have become too rigid, and overlay solutions result in complex management. Hence, the future Internet network is evolving as a complete new concept, which segregates the network layers (Infrastructure, Resources, and Application), allowing for virtualization and software-defined networks (SDN) to be applied.

To redesign such a network, a new Network Operating System (NOS) is standardized, to enable the "network of networks" to interact at several functional levels—processing, storage, communications, and sensing, and to optimize for both the local, internal sphere, and for the global sphere.

Such a network needs to consider differentiated management of the data that flows through it, for consumers, business, and Content Provider. Those who provide effective information retrieval will have a competitive advantage. Such a network must be cognitive—with self-configuring capabilities that respond to highly dynamic changes. It must be programmable, so more IT concepts are incorporated and greater variety of vendors are accommodated.

4.1 Introduction

The main purpose of this chapter is to present some possible paths for the evolution/transformation of the network. It is clear that the new network will have to cope with a gigantic increase in data communication demands. Cisco recently came out with "dramatic" forecasts on the increase of bandwidth request (CISCO 2012), that already have been reduced during the course of 2013. In order to keep pace with this surging demand, both flexibility and cost reductions in the network equipment are required. It is not economically sustainable to continue to invest large amounts of money in technologies that are not guaranteed to be valid for years

R. Minerva and N. Crespi, *Networks and New Services: A Complete Story*,
Internet of Things, DOI 10.1007/978-3-319-33995-5_4

to come. Instead, this evolving environment calls for networks based on low cost technologies that have a Return on Investment (ROI), of only a few years. This is clearly in sharp contrast to the investments in access networks and in infrastructure that Telcos are faced with over the next few years. However, access networks constitute a sort of monopoly and, despite their huge need for capital investment need, they can leverage this advantage. Control and service infrastructure, meanwhile, are more characterized by software and, as such, they are prone to be improved, modified, extended, and replaced in shorter cycles. There is an increasing discrepancy between the evolution of the network infrastructure and the service layer.

The network evolution has always been regulated by standards, and by their progressive and smooth adoption. Operators have had time to align with a controlled and predictable way of evolving along with the network. However, the increasing demand for data (especially mobile data) has dramatically altered this situation. The proliferation of "half generation" solutions like 2.5G, or 3.5G are examples of how the market is forcing operators to deploy more and more capable solutions. Keeping aligned with the evolution of demand and standardization is not easy and it is becoming more difficult. Standards take time to be specified and implemented. Surprisingly, the definition and adoption of new technologies in the access area is faster than in the core (and the service realm). For instance, the LTE (Long Term Evolution) specification is more advanced in the access realm than in the solutions proposed for service control (i.e., the IP Multimedia Subsystem). There is an increasing risk of deploying solutions and architectures that were designed several years ago and which are not able to cope with current network needs and problems.

In addition, there is an increasingly important issue of Internet ossification, i.e., the Internet mechanisms that have been solid for years are now becoming obsolete and need to be updated, both from a technological and an architectural perspective. This section discusses Future Internet requirements and needs, based on current research initiatives [e.g., GENI (Peterson and Wroclawski 2007), Fire (Gavras et al. 2007), Akari (Aoyama 2009)]. These projects propose an extremely flat and programmable network infrastructure that exceeds the concept of communication-only with the integration of distributed processing, storage, and sensing/actuation (or cloud computing) within a common infrastructure.

Driven by the Future Internet, the evolution of the network toward a more software-driven approach based on general purpose systems is capturing the interest of many Telcos. These technologies appear to offer a means to reduce investment costs and allow the Telcos to stay aligned with the technological evolution cycles. Three major trends will be considered, namely Virtualization, Software-Defined Networking, and Information-Centric Networking. They present features that can be implemented in a mainstream fashion (i.e., aligned to the foreseen and standardized evolution of the network) or they can be used in much more disruptive ways. These approaches will be further discussed in this chapter.

In the last part of this section, a few divergent trends are considered, with special attention to their disruptive aspects. These trends are presented to show how future networks could be extremely different from those produced by the linear progression proposed by the standardization process that has thus far described the infrastructure. This analysis shows how trends introduce and use technologies in a different way, creating new possibilities.

4.2 Ossification of the Internet, and the Future Internet

The current Internet is based on autonomous systems that decide internally how to allocate communication resources (in terms of routing capabilities); these resources offer the best-effort communication capabilities. Figure 4.1 represents the independent networked functionalities that work separately without any possible strict coordination, using only the best-effort approach. All of the issues and problems of delay or information loss, as well as security and mobility, are to be solved by the end nodes and the end applications.

A few issues are related to IP ossification. As stated by Turner and McKeown (2008): "*Today, many aspects appear to be set in stone, such as difficulty in getting IP multicast deployed, major obstacles to deployment of IPv6, not optimized routes (problems with BGP), extensions to routing—such as multicast, anycast, ...*". This poses the need to change the Internet to allow for natively encompassing security, mobility, better resource management, and pervasiveness features. It is quite difficult to make any changes in the infrastructure of the Internet, and following the

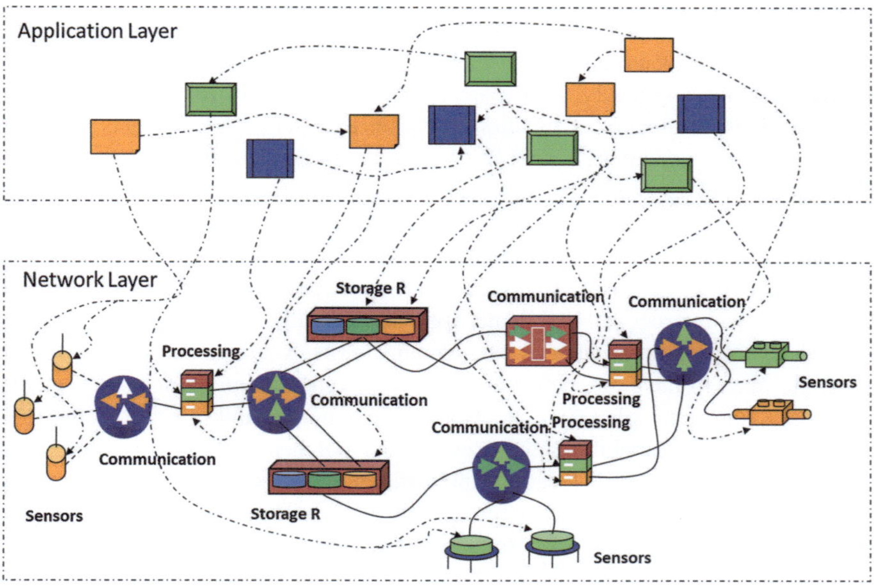

Fig. 4.1 Independent autonomous systems supporting best-effort capabilities

end-to-end principle (Clark et al. 2005), major changes are possible only at end nodes, not in routers. In addition, this monolithic approach has been exploited by a small number of companies to create a sort of monopoly on the IP layer based on closed routers and systems. As a consequence, more flexible solutions have been implemented by means of overlaying mechanisms (i.e., building virtual networks on top of the Internet, e.g., P2P). These overlay networks clearly show programmability and flexibility advantages, as well as the possibility to improve the underlying protocols. Some examples are the JXTA P2P architecture (Wilson 2002) or many solutions created on the BitTorrent protocol (Pouwelse et al. 2005).

The major Internet rigidities are as follows:

- Security: the current Internet has not been designed to cope with the security issues caused by the availability of a multitude of distributed computers. Classical examples of Internet flaws in this are the Distributed Denial of Service (DDOS) attacks that exploit the possibility of different sources to concentrate traffic and request on a single or a few server nodes, causing these to go out of service. The network as a system does not have the capabilities to detect and act against these and other threats.
- Availability/Reliability: having adopted an autonomous system approach with limited cooperation between nodes and adjacent subsystem, the reliability of the solution resides in the capability to route information over multiple alternate paths. There is no assurance that a system will be up and running and whether it will provide a certain level of service. The entire gamuts of services developed over the Internet are unreliable by definition.
- Predictability: systems offer neither the means to determine or control whether a path or a set of subsystems will be available, nor if a service can be provided in a predictable way. There is no possibility to predict beforehand how a service or a subsystem will behave.
- Manageability: typically, nodes are closed systems with limited management capabilities. Management functions are exerted essentially by means of Client Line Management (CLI). This requires a considerable amount of work from human operators and often, important management functions must be executed manually. A big source of routing problems is related to human-based node configuration.
- Mobility: in an increasingly mobile world with accessible mobile terminals and devices, the Internet and mobile networks are not fully supporting the possibility to access objects and functions on a dynamic basis. Roaming and its issues, such as the need to have an anchor network (the home network) for accessing services is one example. Another example is the availability of a plethora of communication capabilities and devices but not being able to use and share them. The problems of locating applications and objects, as well as creating local supporting environments for optimizing user experience are the concepts that are out of the scope of current networks.

- Sensing: sensors are not very well-integrated in the Internet (some solutions have even proposed moving to different protocol than IP (Wan et al. 2002; Dunkels et al. 2003), because it is cumbersome and over specified for small devices). Sensor networks will also need different types of transport capabilities and they will in all likely benefit from the reliability and predictability of network services to support mission-critical applications.
- Scalability: large-scale systems are difficult to maintain and operate over the Internet. Mechanisms and solutions are needed to facilitate the integration of different components. This will become even more acute since the future network will not provide simple connectivity, but will evolve toward the integration of communication, processing, storage, and sensing/actuation. These features are neither integrated nor natively supported by the Internet.

The Internet is a major building block of an increasingly critical infrastructure for people, agencies, and companies. The features detailed above need to be present in any new infrastructure in order to improve the effectiveness of the Internet. Those features also point to persistent issues that cannot and have not been solved with incremental improvements to the current architecture (Roberts 2009; Rexford and Dovrolis 2010). There is much work invested in developing out new, more powerful paradigms on which to build the so-called Future Internet.

Another important issue that is not dealt within the current Internet is the clear separation between the "network" stratum and the application one. In principle, this separation is valuable, but it creates problems because the network cannot adapt to or fulfill the dynamic requirements of applications. A level of interfacing between the network and an application could guarantee that the resources of the network are not wasted, but rather used cost-effectively to reach two optimization goals: (a) the usage of resources on the application side, i.e., the resources are allocated to fully satisfy the needs of the application; and (b) the resources are optimized at the global level, i.e., the resources are allocated as they are required to fulfill the needs of competing applications and a fair strategy is applied to minimize the use and allocation of resources to applications. This twofold problem needs to be supported by the Future Internet to solve some of the previous issues (e.g., manageability and predictability).

With respect to Fig. 4.1, there is a need to introduce a higher level of cooperation between systems and nodes so as to offer a better quality service (it is not the case to introduce guaranteed levels of services, but instead to guarantee that resources are utilized in such a way that the solution adopted can be considered "quasi-optimal" and that during time (if conditions stay stable) the solution can converge versus an optimal one. Simple algorithms and protocols can be used with the purpose of allowing changes in status of resources so that they can progressively find optimal solutions. Jelasity et al. (2005) demonstrated that a complex system can converge toward the optimal solution under this condition. Figure 4.2 illustrates the difference between a best-effort approach and a cooperative one.

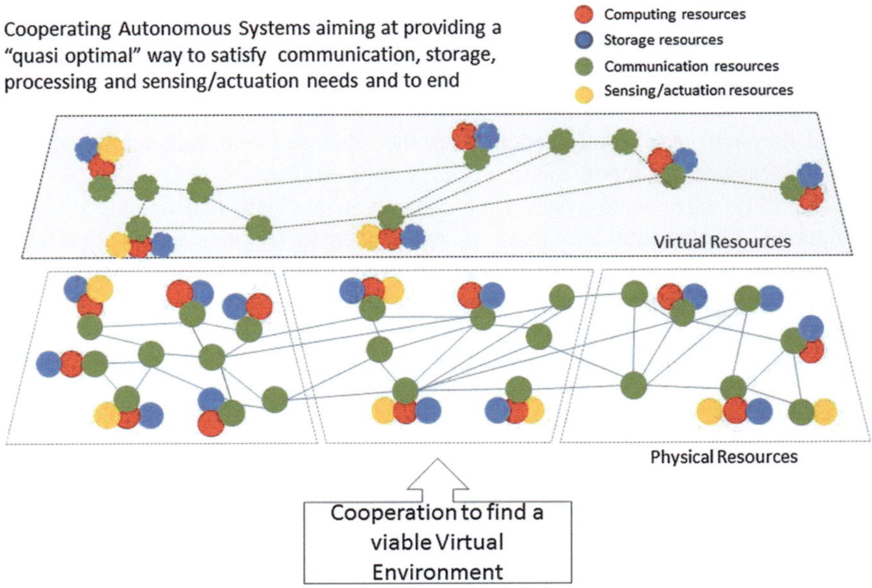

Fig. 4.2 Cooperative systems in which each node has a simple behavior

Two major properties are needed with the purpose of further innovating the Internet Architecture:

- Programmability in the nodes: if programmable functions are instantiated in Internet nodes, a high level of network flexibility and adaptability can be reached. Programmable nodes are the building blocks that allow to make the Internet less monolithic and are the key to introduce richer functions and features in it.
- In order to keep the pace with the solutions and the mechanisms needed to program and control a complex infrastructure, virtualization could be widely used. It brings the ability to colocate multiple instances of network services on the same hardware—each running in a different virtual machine, but also the capability to confine specific applications in a virtual machine. Crashes and security issues of an application can be confined to a specific virtual machine. In addition, the set of virtualized applications and functionalities can be increased depending on the demand of users by replicating virtual machines in the system. So virtualization seems to be a useful mechanism to move toward the Future Internet.

This path to change could be interpreted in different ways, some initiatives like GENI (Peterson and Wroclawski 2007), Akari (Aoyama 2009), or FIRE (Gavras et al. 2007) are adopting a clean slate approach, i.e., they advocate the need of a totally new approach for defining the Future Internet. Many Vendors, instead, are more conservative in this approach and they propose the progressive opening up of routing interfaces to the application layer. These two trends will be further discussed in the next sections.

4.3 Virtualized Networks

Virtualization has widely been exploited by the IT industry, a working definition of Virtualization is "the ability to run multiple operating systems on a single physical system and share the underlying hardware resources" (VMWare 2006). Virtualization carries in a number of advantages such as:

- Expanding hardware capabilities, allowing each single machine to do more simultaneous work
- Control of costs and simplification of management through consolidation of servers
- Higher level of control of large multiprocessor and cluster installations, for example in server farms
- Improvement of security, reliability, and device independence possible thanks to hypervisor architectures
- Ability to run complex, OS-dependent applications in different hardware or OS environments.

But why and how to approach virtualization in the Network? Today there are already some low levels of virtualization of the network. For examples, solutions for managing and virtualize connectivity have introduced functionalities related to the creation and support of virtual networks based on MPLS or GMPLS and VPNs. This virtualization, however, offers a coarse-grained network link virtualization that is far from allowing the management of fully fledged virtual networks. Examples of virtualized networks are overlays networks (as P2P networks over the Internet). They can be seen as virtual networks at the application level. One issue is the relations between the virtual network and the underlying physical infrastructure. This missing nexus is highly detrimental either at the application level or to the physical level, in fact applications cannot control how networked resources can be allocated and used, while the physical network resources are not optimized because the overlay network can refer and use resources that are far from the actual location. The issue [tackled by P4P (Xie et al. 2008) and ALTO (Gurbani et al. 2009)] is that the separation between applications and the network is not providing optimized solutions, resources are misused and the service returned is not always adequate for the applications.

A network wide virtualization (using the same paradigm used for IT resources) would allow:

- To optimize the use of physical resources (as previously discussed).
- To integrate deeply IT and Net resources in virtual networks tailored to applications requirements, this is a giant step with respect to actual situation just focusing on the connectivity.
- To operate independent virtual networks "dedicated" to different Users and migrate them when necessary. So applications and services can be segmented and can use specialized networks to reach their goals. One example is the ability to create independent network slices "dedicated" to different Players (e.g., virtual network operators, application service providers for video streaming, Content Delivery Networks (CDN), etc.).

Recently, a group of Operators has started a new initiative in ETSI called the Network Function Virtualization (NFV), whose goal is to foster the wide adoption of virtualization within the network. The grand plan is to create a general purpose computing infrastructure on top of which to instantiate several images of specific nodes. This has the advantage of decoupling the hardware procurement from the software one. In addition, instances of specific nodes could be acquired by different vendors, and deployed remotely by using standard IT solutions. Figure 4.3 represents these possibilities.

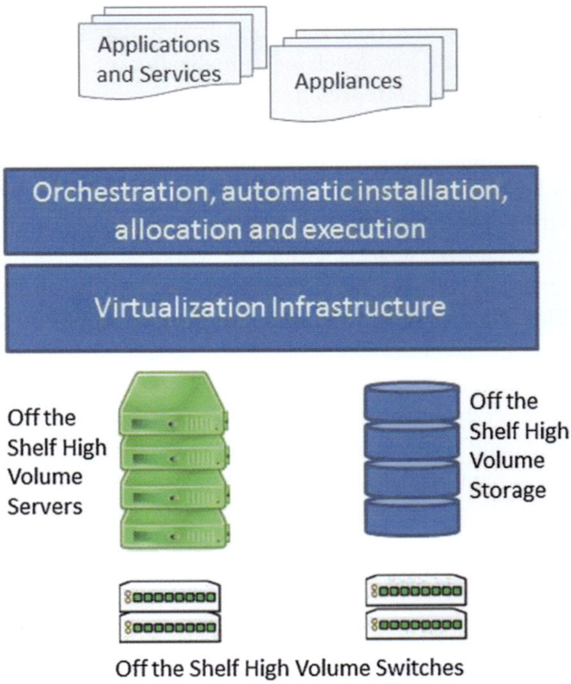

Fig. 4.3 Network virtualization approach

4.4 Toward Software-Defined Networks

Together with the Virtualization, another trend is acquiring momentum in the industry: the software-defined networking was spawn in 2008 by studies carried out in Stanford for a clean slate approach toward the evolution of the Internet (Yap et al. 2010; Koponen et al. 2011). This approach draws its root into older approaches such as programmable networks (Campbell et al. 1999) or even TINA (Berndt and Minerva 1995).

In traditional IP networks, a node has its own control plane and the management/policy plane to actuate configurations with CLI (Command Line Interface), there is an automatic support for dealing with the complexity of a whole system and avoiding configuration issues. This is also a major source of errors and issues in the network. In Software-Defined Network, control and data planes are decoupled, so network control and states are logically centralized, and the underlying network infrastructure is abstracted from the applications. This means that the routing nodes can be controlled by means of specific protocols such as OpenFlow (McKeown et al. 2008) and all the intelligence can be moved somewhere else. In addition, the switching capabilities can be exerted by very low cost devices that can be bought for low prices. In this way, the Control Level becomes programmable because SDN offers programmable interfaces (i.e., APIs) to the network. It is then possible to implement new "routing" protocols (e.g., customize paths for network traffic engineering), to intelligently allocate resources to the needed applications. The rich and intelligent functions move to a programmable infrastructure that changes the way in which networks (and autonomous systems) are governed and orchestrated. Network-wide policies specific for customers or services can be implemented and controlled. The network can be controlled and managed with a holistic approach. This means that cognitive capabilities can be applied, and the network will become highly adaptive and responsive to changes in its usage.

The next step is to create a NOS (Fig. 4.4), i.e., a set of functions and interfaces that programs can use so as to access the services of the infrastructure. The consequence is that not only the routing mechanisms, but also multicast, processing, and computing became services that can be mashed up and composed. The definition of a NOS will allow applications to orchestrate how the network responds to the requests of the applications. The rigid separation between the network level and the application one is reconciled by means of APIs and programmability.

This trend will have a deep impact on the industry and IP protocols. Operators willing to provide services and platforms have to consider the availability of these platforms. Operators interested in being Bit Carrier, will prefer to remain at level L1–L2 and to exploit the optical technologies in order to save on energy, processing, and intelligence in their network. This is one of the most important inflection points (and decision points) in the evolution of technology, because it shows that large networks with a considerable bandwidth can be managed and operated by functions of L1–L2. Moving at upper layers will cost money because systems are more intelligent (and more expensive) and they consume more power.

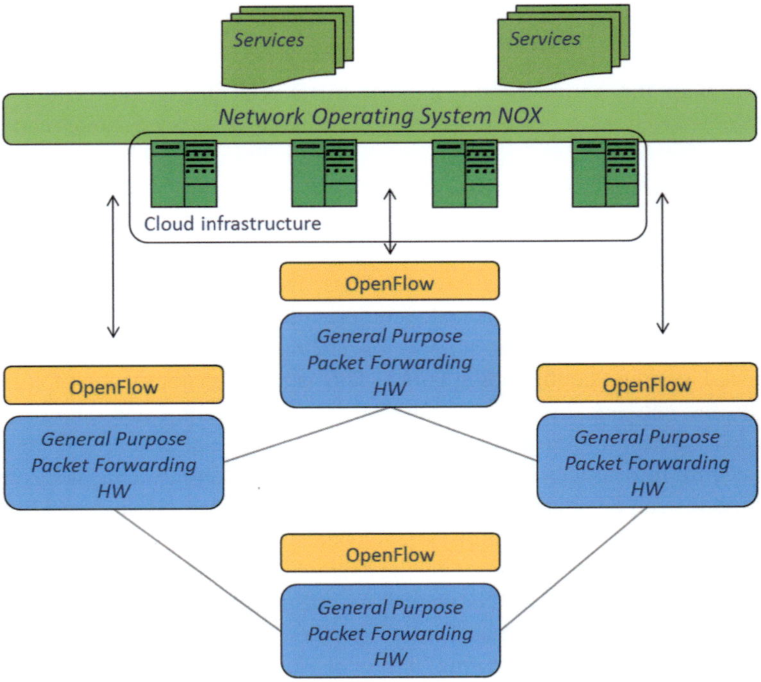

Fig. 4.4 A network operating system, NOX, based on OpenFlow interfaces

A Bit Carrier will try to use as much as possible lower level equipment and it will avoid to move to upper levels. With the purpose of fully appreciating the innovation carried by the software-defined networking, an example is explicative. Using the concept of NOX, an Operator A can offer to Operator B the usage of its resources by means of open Interfaces. Operator B could even decide that the most important pieces of logic will remain remotely located in its data center, but the effect of control capabilities will be exerted to the local level (to the OpenFlow-enabled resource). Roaming could be totally reshaped with this approach, as well as the relationships between the Operators and other infrastructure providers. In fact, a company can offer its programmable resources to Operators that can implement on top of it services for their clients. Figure 4.5 tries to represent a few of these capabilities.

This example shows how programmability and opening up of interfaces of resources leads to deperimeterization of services and networks. Actually, a provider could negotiate dynamically the allocation of needed resources, independently from the location, and to build a solution to support the connectivity and the services of its customers. Services are not strongly tied to the physical network infrastructure anymore. Perimeterization is a major drawback that operators have to solve so as to compete worldwide. The consequences and the business models behind this new approach have still to be fully understood and explored.

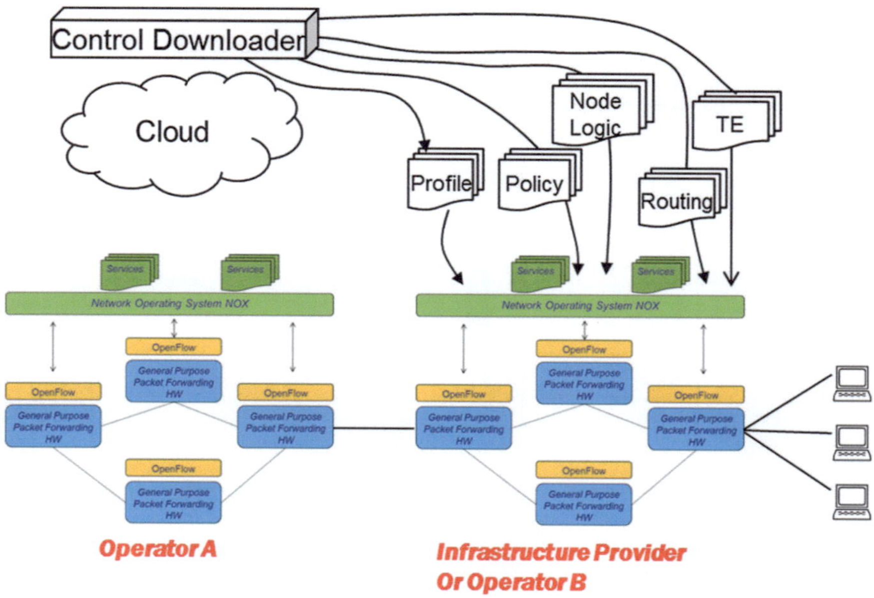

Fig. 4.5 Deperimeterization of networks by means of SDN

4.5 Information Centric Networks

In the context of Future Internet studies, there is a trend focusing on the importance of accessing the data and information. In a seminal presentation, Jacobson et al. (2009) presented the concept and justified it by saying that over 90 % of communication was human to data (i.e., a human accessing to data) instead of a human–to-human (or a point–to-point) communication. Adopting a data-oriented view leads to the idea to reshape the networking capabilities focusing on how to access the indexed data. Data should be identified and stored in a sort of large P2P network and information accessed according to several policies established by the owner of the information and the preferences and rights of the user. From an architectural view (Rothenberg 2008), this approach could lead to a three layer functional architecture. At the lower level, a data plane for the control and transport of chunks of data and information. At an intermediate level, an information plane is in charge of dealing with the information, i.e., naming and addressing, indexing of content, management of replication, mediation of formats, data mining and correlation, reasoning and inferring of information from available data. At the upper level, the functions are related to how data are accessed and made available to users, to security of data. Another important set of functions could be related to how the data can be certified, updated, and even canceled. This aims at lessening the problem of not authorized duplication of data. At this level, solutions like Vanish (Geambasu et al. 2009), that allows the users to have control on data that they "post" on the

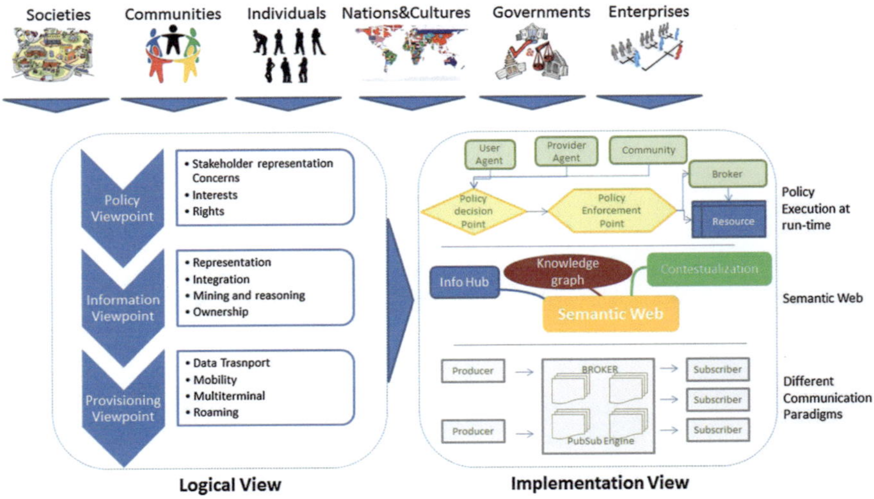

Fig. 4.6 Architectural vision and architectural implementation (*Source* BT)

Internet, could find a natural context for usage. In addition, this layer could be in charge of dealing with how providers can control and establish policies for usage. On top of this infrastructure, Application Programming Interfaces (APIs) could be provided in such a way to allow the creation of applications or simply the management of data. Figure 4.6 depicts a possible architectural view of the ICN considering these different plans.

From the implementation point of view, the architecture can be based on a number of available technologies: at the lower level a P2P infrastructure for data exchange (e.g., based on BitTorrent) could take care of the optimization of the bit exchange, while a PubSub (Fontoura et al. 2013) infrastructure could help in notifying the availability of new data or duplication or update of existing data. With the purpose of giving a glimpse of possible solutions, an architecture could be organized as follows:

- At the information Plane level, a brokering infrastructure could find a viable application, for instance an infrastructure based on Distributed Hash Table (and the federation of them) could help in tracking the available data and information, also NoSQL solutions could be used to store indexing and pointers to data objects.
- Data mining solutions could also be integrated at this level.
- At the upper layer, there is a need to implement a sort of Policy-Based Systems (Follows and Straeten 1999). At this level, the Policy Decision Points will decide and determine how to grant access and the permitted usage of Data while at the underlying levels, Policy Decision Points will retrieve, format, and present the data accordingly.

The topic of “Information Centric Networks” had a moment of hype in the past years and currently it is not so popular. However, it points to meaningful and important issues of New Networks that deserve research and thinking. At least these aspects are relevant:

- The need of highly distributed architecture for dealing with information storage and replication and fast delivery to users (CDN).
- Optimize architecture for information retrieval especially in relation to the personal data issue.
- The issues related to the policies of accessing data and information (e.g., the mentioned Vanish approach).

Current CDN architectures are a sort of overlay network not integrated within the networking infrastructure. In future, the network will be a programmable combination of processing, storage, communications, and sensing capabilities. The ideal case is to have solutions that can integrate the SDN capabilities with intelligent caching and processing. Actually, new mechanisms for sharing algorithms and programs that produce the wanted data as a result of calculation (Katti et al. 2008) can be considered as an alternative to transport an overwhelming quantity of data. Transferring a program and initial data could be more efficient than transporting the entire mass of data. The user processing capabilities can be used to execute the algorithm for deriving the desired data. Networks could be totally reshaped by similar approaches.

From an Operator’s point of view, a possible strategy for dealing profitably with data could be the one to optimize the resources for transferring the bits (at the data layer) and to develop differentiated mechanisms for dealing with personal data (consumer market), company data (business market) and Content Provider data (e.g., the big web companies). In each case, there is the possibility to optimize the level of experience of the customer. The underlying need of different customer segments is the same: data are generated as results of processes. They can be personal processes (e.g., shopping, traveling, community experience) that can be made evident to the user so as to help him to understand his behavior; or they can be company processes (how a company deals with the data produced by its customers, data related to internal processes and the like); or they can be related to how data flows are generated and consumed by customers of content providers. Being able to optimize how data are managed and handled in an enterprise means to understand the internal business processes. Many companies have the need to improve processes and to save money by making those processes more effective. Acquiring this knowledge is a change of perspective that gives new opportunities to the Operators. This is another inflection point: winning companies (and Operators) will be able to intercept data and help the customers to improve their internal and external processes.

4.6 Cognitive Networks

Many technologies fall under the umbrella of "Cognitive Networks." They range from Software-Defined Radio (Jondral 2005) to Autonomic Computing (Marinescu and Kroger 2007). The communality is the attempt to introduce higher levels of intelligence in order to optimize the usage of complex resources. (Manzalini et al. 2012) presents an extensive treatment of the technologies, and their evolution. The interest and the trend toward cognitive networks is consolidating at the standardization level [e.g., the AFI initiative (Chaparadza et al. 2009)] and within the industry [for instance the new NoOps approach (Fu et al. 2010) aiming at leveraging automation, from development through deployment and beyond, to increase the speed at which applications can be released in a cloud environment]. The cognitive capabilities are considered under two perspectives: 0-touch networks (Manzalini et al. 2011a, b) (or systems), i.e., how self-configuring capabilities can help in coping with highly dynamic changes in configurations of networks and systems; Networks of Networks, i.e., how intelligent functions can be used so as to allow the self-organization of complex systems (Manzalini et al. 2010a, b, c, d). Future Networks will integrate at least processing, storage, communications, and sensing capabilities provided by highly distributed systems. These systems will be highly heterogeneous and they will pertain to several administrative domains. Management will result in more and more complexity, because the systems will need to integrate functions that have been separated so far. Indeed, the processing and storage from communications, sensing, and actuating are three different functional groups with their own specific requirements. The single resource will be forged by the integration of hardware (proprietary and general purpose) and basic and specific software (possibly by different providers). In addition, systems will depend on different management authorities, each one with its policies and management requirements. Another important point is that systems will aggregate a work together in a very dynamic way: the concept of ephemeral networking makes this concept very explicit (Manzalini et al. 2012a, b). For these reasons, it will be extremely important to approach the problem of highly distributed and heterogeneous systems in a different way: cognitive technologies could be extremely helpful in changing the way of how systems are managed. Each single resource should be able to detect its own behavior and optimize it with respect to its intended working and to the environment in which it is integrated. This involves two control loops used to optimize the behavior of the resource: a local one, intended to optimize the internal behavior and a global one, used to optimize the behavior of the resource with respect to the external environment. Figure 4.7 depicts this concept.

In order to devote the large part of processing capabilities to the functional goal of a resource, the approach should favor the lightness of the management capabilities. Under this respect, gossiping-based solutions seem to be attractive because they reduce the "behavioral load" of the resource. In addition, it has been demonstrated that gossiping algorithms converge to the optimal solutions, if enough time and iterations between objects are given (Jelasity et al. 2005). This is important

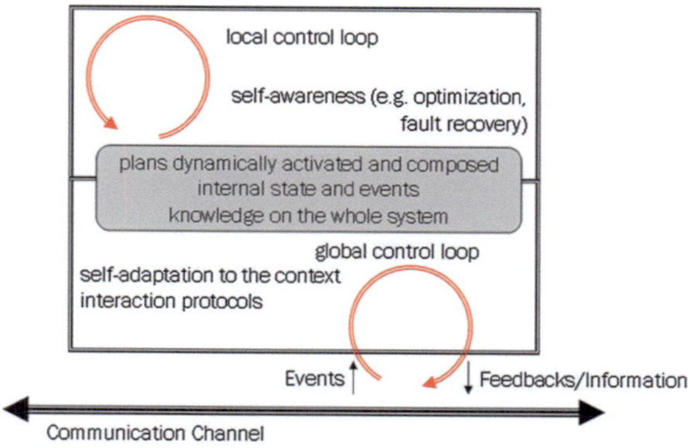

Fig. 4.7 Local and global control loop

because simplicity can help to converge towards optimal solutions. However, a system made out of many cooperating resources (and not all of them are reliable) is changing very frequently and computing the optimal solutions cannot be practical or even useful. So, a few iterations of the system can lead to a refinement of the solution (a quasi-optimal one) that is good enough for the system to save processing time. A number of optimization algorithms and solutions have been studied in Manzalini et al. (2010, 2011a, b), Moiso et al. (2010).

Another important feature of autonomic systems is that they can contribute to make related systems more reliable. In Manzalini et al. (2010), a situation in which a set of inner autonomic resources use their features to "enforce" some levels of autonomics to unreliable resources has been proposed. A similar situation can be depicted as in Fig. 4.8

The autonomic reliable resources can be used as a sort of reliable set of resources that unreliable resources (at the crown of the circle) can use to store, process, and integrate data in an intelligent and robust way.

Figure 4.9 depicts three cases in which physical resources cooperate by means of gossiping protocols so as to ease the supervising and orchestration of resources for upper layers. A virtual resource can have autonomic properties [case (a) and (b)] or can be unreliable [case (c)].

An Inner Circle could be composed of peers managed by different administrative domains, provided that they (demonstrate to or are certified to) offer suitable levels of reliability and trustiness. In a certain sense, the Inner Circle properties can be such that they can be inherited (or exported towards) external entities and environments [see Fig. 4.9 case (b)]. For example, the supervision features implemented on the peers of the Inner Circle can be used to export these properties toward applications which use also resources provided by other domains that do not show the same stability and reliability (e.g., nodes are executed in a highly unstable environment, peers are not completely reliable and trusted and their functions should be carefully used, and the like): they can be used to enforce reliability policies and to identify contingency plan in case of need. Figure 4.10 represents the aggregation and orchestration of (virtual) resources in order to fulfill nonfunctional

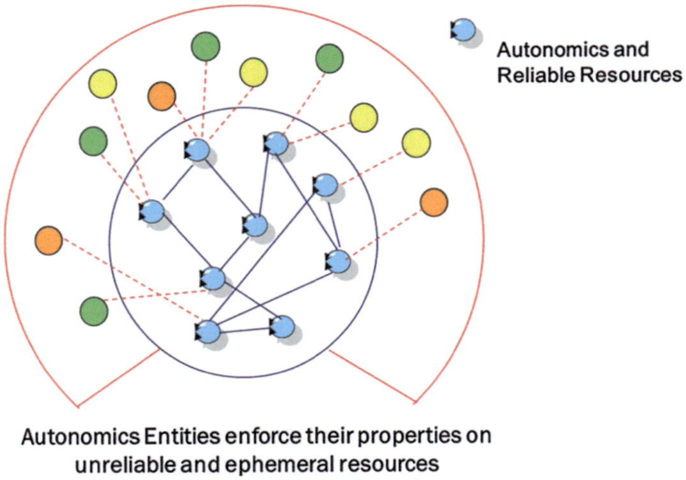

Fig. 4.8 An inner circle of reliable resources

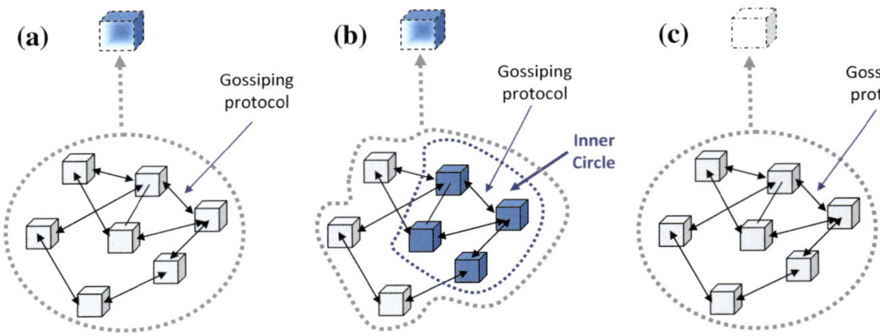

Fig. 4.9 Creating autonomic virtual resources. **a** Unreliable resources cooperating to support a reliable automonic virtual resource. **b** Unreliable and reliable (*blue*) resources cooperating to support a reliable autonomic virtual resource. **c** Unreliable resources cooperating to support an unreliable virtual resource

requirements at the level of virtualized autonomic overlay of resources. The Inner Circle concept is "recursive"; it can be adopted at several layers of the architecture. This property guarantees the openness of the architecture and the possibility to integrate different types of resources from different domains (deperimeterization).

Operators and Players aiming at providing services can exploit the Inner Circle in order to allocate (virtual) resources for the creation of a reliable and trusted distributed execution and networking environment. They are specialized through the deployment of software modules for service logic, logic for controlling objects, and DB/repository schema, and interconnecting them according to the needed communication patterns. In particular, the resources can be aggregated according to a distributed architecture able to deliver applications according to the control

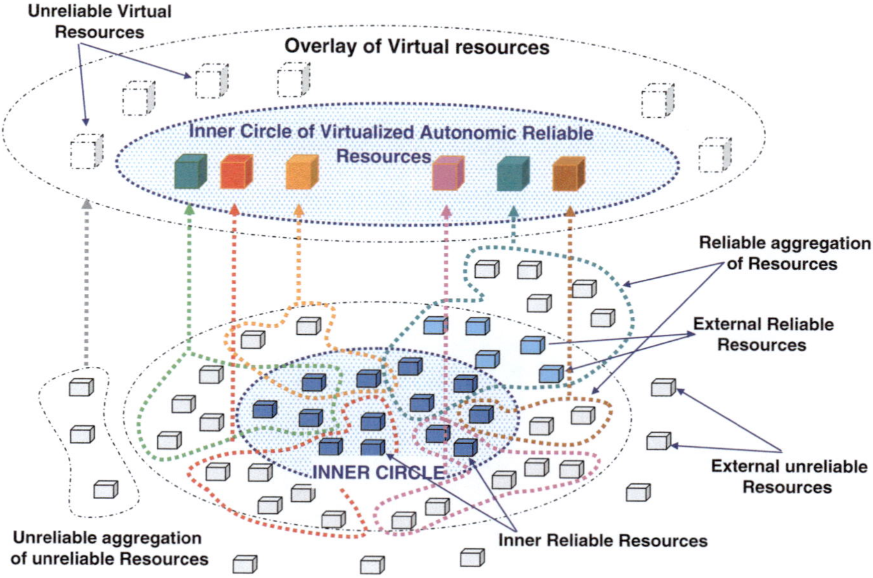

Fig. 4.10 Inner circle at the physical and resource aware levels

paradigm most suited for each of them. This will also guarantee the availability of a minimal set of always available resources that will support a minimal set of vital functions for services and applications.

4.6.1 General Principles

The proposed architecture (Fig. 4.11) relies on a small set of principles shaping the envisaged structure.

Entities and related functions are organized in three layers: **Infrastructure Layer** comprises entities that provide a virtualized view for communication, storage, processing, and sensing/actuation (things) resources; the **Resource Aware Layer** comprises entities and functions needed to supervise and optimize the usage of the virtual resources; **the Application Layer** is made out of applications that use the resources of the Infrastructure. All the entities in Cognitive Service Platform are represented as *resources*, i.e., entities which provide capabilities and can be allocated to/used by applications. Different types of resources can be envisaged. For instance, *infrastructural resources* provide features (implemented by physical entities, such as servers, communication network nodes, and machines) to create distributed execution/networking environments for deploying and delivering services; *application resources* provide logical functions/data to be composed and aggregated for creating applications (or other composed application resources); they

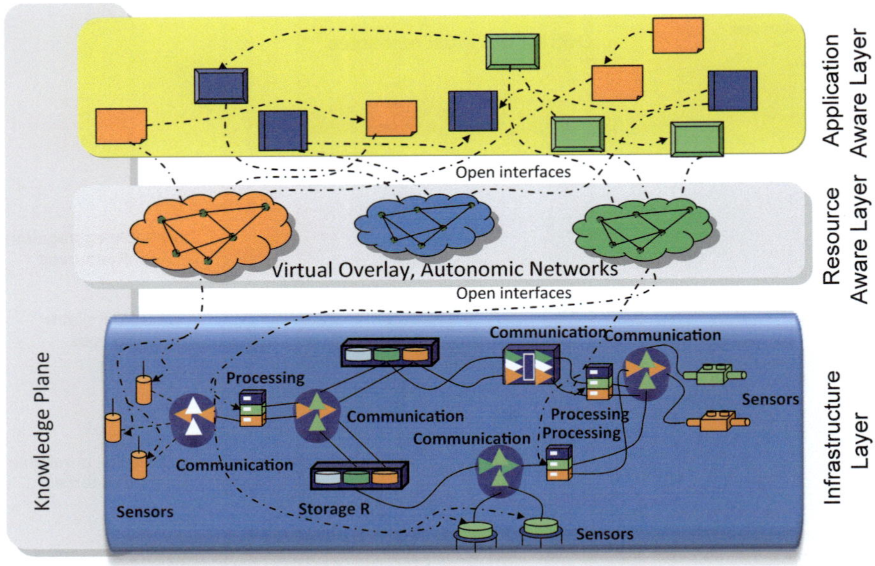

Fig. 4.11 A system view of cognitive service platform

could rely on capabilities of infrastructural resources allocated to them. Capabilities of physical and logical resources are virtualized with the purpose of ensuring secure, personalized, and isolated usage. Virtualization includes features for abstracting (e.g., simplifying, copying with heterogeneity) the interfaces to access/use resource capabilities, for sharing them (e.g., by partitioning their capabilities in isolated portions singularly allocable), and for providing a formalism for their description (e.g., to be used by allocation/negotiation functions). Each resource type can adopt a specific *virtualization paradigm*. The behavior of the (virtualized) resources is enriched with *autonomic features*: self-awareness and self-organization capabilities (e.g., by adopting control-loops mechanisms), that make the entire infrastructure more robust, controllable, and resilient. Autonomic resources can self-adapt their behavior to achieve self-management, to react to internal/external events, and coordinate with other (nearby) resources. As demonstrated in (Jelasity et al. 2005), a global behavior emerges from local decisions and interactions (a typical mechanism adopted for the governance of complex adaptive systems).

The (virtualized) resources are clustered into *overlay networks*, used for the exchange of messages among the resources (which have to implement logic for joining, maintaining, and possibly optimizing the overlays to which have to participate). Overlay networks implement several control and supervision algorithms, such as resource discovery, resource allocation, load balancing, fault detection and

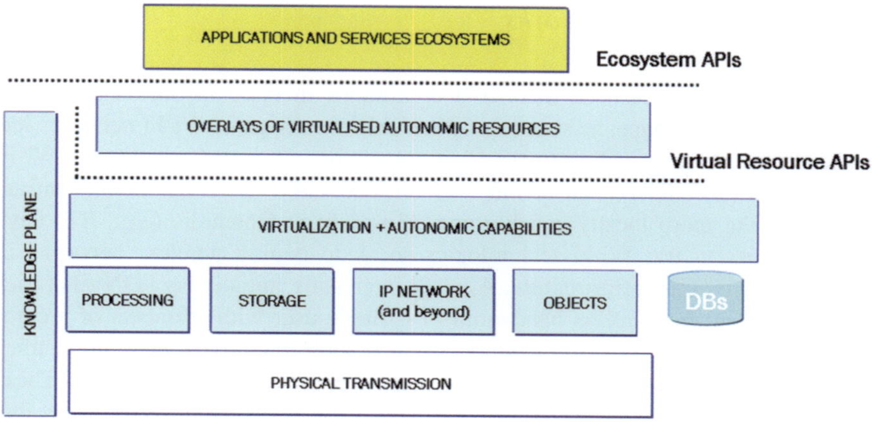

Fig. 4.12 A layered view of cognitive service platform

recovery, dissemination of system information. The protocols adopted for their implementation are based on *gossiping*. They are based on iterative information exchange: during each protocol step, a node exchanges with (a small subset of) its neighbors in the overlay a small amount of data, and combine them to update its local state. In order to fully orchestrate a complex infrastructure that collects several networks and a multitude of end points, cognitive approach to infrastructure governance is undertaken. A Knowledge Plane (KP) is introduced aiming at handling the available cross-layer knowledge gathered from all the protocols being used and from all the resources (and their virtualized representation). The KP is fundamental for effectively creating virtualized systems that fully satisfy Users and applications requirements. These principles are embodied in the layered architecture depicted in Fig. 4.12. On top of this infrastructure, applications and services can be built and combined by aggregating the virtual resources and extending them by introducing new resources. This enables the creation of an ecosystem where several players (e.g., Individuals, Enterprises, Service and Network Providers) can interact in an open and collaborative way to produce/consume/sell/buy/trade data, contents functions, and applications.

4.7 Disruptive Approaches

There are plenty of activities aiming at redefining and implementing new solutions for the network architecture. Here, some of them are sketched out to give a glimpse of the underlying complexity and richness of research around the theme of new networks.

4.7.1 Local Area Networks

They are important because there is a continuous increase in the usage of this technology. For instance, as stated in GIGAOM,[1] the usage of Wi-Fi networks has exceeded the usage of cellular nets. Many Operators are trying to integrate Wi-Fi Networks with cellular ones in order to off-load traffic from the mobile infrastructure to the more locally capable ones. Some fixed Operators (e.g., BT) have tried for long to use Wi-Fi technologies so as to deploy wireless networks in alternative to mobile infrastructure. Another interesting initiative is FON[2] that tries to share part of the wireless bandwidth of a single user with members of a community (in this case the FON network). Behind these initiatives, there are studies and experimentations aiming at creating and supporting mesh networking. These initiatives are based on the possibility of dealing with access points as resources that can be hacked and programmed. Indeed, there are some distributions of software [e.g., openWRT (Fainelli 2008) and DD-WRT (Weiss 2006)] that can be deployed on existing access points making them programmable and configurable according to the needs of the users. For instance FON is based on openWRT. Along this mesh networking studies, the need to optimize the routing between mesh nodes has brought to develop new routing mechanisms capable of coping with the dynamicity of mesh systems. One interesting protocol (that anticipate somehow the needs of the mentioned autonomic systems) is the B.A.T.M.A.N. protocol (Johnson et al. 2008) that uses a sort of gossiping mechanism in order to create routing paths within mesh networks. Another interesting experiment is Netsukuku (Lo Pumo 2007) that aims at creating a network without any central authority and not controlled by an ISP. These two experiments are particularly interesting because they have leveraged the availability of open source for access points so as to create new community networks that have a great disruptive potential. Studies in Wi-Fi and local area networks are then very important for radical alternative to public and strictly controlled networks.

4.7.2 Community Networks

"Home with Tails" (Slater and Wu 2009) and the seminal work of Arnaud et al. (2003) and more recently on Free Fiber to the Homes[3] propose a very disruptive model for the access network: the user is the owner of the fiber connecting the home to the cabinet (and maybe even to the central switch). This simple proposition is strongly User Centric (and hence the interest in it) and it has a very high disruptive

[1] http://gigaom.com/2012/07/10/we-already-use-wi-fi-more-than-cellular-why-not-continue-the-trend/. Last accessed April 10, 2013.

[2] http://corp.fon.com/. Last accessed April 10, 2013.

[3] Available at http://free-fiber-to-the-home.blogspot.it/. Last accessed April 9, 2013.

potential: people can manage their connectivity and can create communities to share the communication capabilities. Within large condominium, the infrastructure is owned by the community and internal communication can be totally free of charge. Communities can even create small data centers so as to support the processing and storage needs of the users.

An interesting and related concept is the one of Open Access Network (Battiti et al. 2003), i.e., a network organization that clearly separates the ownership of the infrastructure from the service provision. An example of this type of network is Guifi.net. As stated in the website[4]: "*guifi.net is a telecommunications network, is open, free and neutral because is built through a Peer-to-Peer agreement where everyone can join the network by providing his connection, and therefore, extending the network and gaining connectivity to all.*"

Another interesting experiment and community is Village Telco (Adeyeye and Gardner-Stephen 2011). It is based on a mesh network made up of Wi-Fi mini-routers combined with an analogue telephone adaptor [aka 'Mesh Potato' (Rowe 2009)]. This infrastructure connects the village to a Session Initiation Protocol (SIP) based server that can be seen as a gateway toward other networks. There is a possibility to use FreeSwitch,[5] i.e., an open-source implementation of a Softswitch, for controlling the connectivity of the local set of SIP phones and the interworking with other networks. In principle, these types of implementation could even make use of openBTS, i.e., an open-source implementation of a GSM (Global System for Mobile Communications) base station (Burgess et al. 2008) for creating an open GSM network.

4.7.3 Military Communication (MilCom)

It comprises several projects aiming at the improvement of communication technologies for military usage. For instance Direcnet[6] is an attempt to build a military communication architecture based on open standards and existing technologies. It is promoting the integration and exploitation of self-configuration with a mesh, sensor, Peer-to-Peer, and cognitive radio networking technologies. This infrastructure creates networks able to provide considerable bandwidth to the single soldier (DirecNet), especially if the number of nodes per area increases. Other projects aim at ensuring communication and data consistency in several Mobile ad hoc Networks (MANET) (e.g., Bastos and Wietgrefe 2010).

These solutions are for military usage now, but as stated for Direcnet, they have the objectives of being open and using COTS systems to reach also civil usage in

---

[4]Available http://guifi.net/en/what_is_guifinet. Last accessed April 9, 2013.

[5]Available at http://www.freeswitch.org/. Last accessed April 9, 2013.

[6]Available at https://www.opengroup.us/direcnet/documents/16322/2012_MILCOM_DirecNet_Technical_Panel.pdf. Last accessed April 9, 2013.

the near future because the envisaged technologies are already available and they just need to be standardized. Using these technologies with a User-Centric approach, i.e., giving the power to the final users, could allow the creation of very powerful networks capable of competing in terms of features and capabilities with the traditional Telco infrastructures.

Direct Communication has the potential to substitute, for some applications, traditional networks exploiting the increasing processing and communication capabilities of new terminals.

4.8 Findings

This Chapter has investigated the possible evolutions of the network. A first remark is that the Internet will go through a substantial change in order to correct and limit the ossification issue. This will bring into the network a lot of interesting functionalities and capabilities like more cooperative approaches between different subsystems and a progressive adoption of virtualization capabilities. In addition, the gap between the network and the application will be closed by means of APIs. This is a long pursued goal, but in the future, it will be finally reached by guaranteeing to application the possibility to adapt the network to the real communications needs of services. Actually, the network itself will not comprise only communications resources anymore. On the contrary, it will encompass processing, storage, and progressively also sensing and actuation elements.

Information-Centric Networking (even if the hype of this approach is over) points to a big problem to be considered by the Telcos: the usage and transport of data is prevalent with respect to the human–to-human communication. This requires specific attention in order to tailor the network to this new prevailing goal. The point of supporting a better "information" retrieval within the network is still a priority in the ICT industry and the companies that will be capable of accommodating this task in an effective way will have a competitive advantage. Virtualization of specialized networks (for instance virtualized CDN) for supporting a specific task could result in new business opportunities.

Software-Defined Networking is a means to control and optimize the usage of available resources. Virtualization is a means to run logical networks on top of a physical infrastructure made out of general purpose machines. Both technologies have a high disruption potential, but when integrated their capability to change the status quo further increases. Some Telcos are reluctant to fully exploit the disruptive potential. In fact, the combination of these two technologies will be used by Telcos in two different fashions (this is drawn on discussions with different Telcos and the analysis of a few proposals of European Projects[7]):

[7]This information cannot be openly disclosed.

- The most innovative Telcos will try to use SDN (and the related Virtualization techniques) to create different networks and to adopt new control paradigms as well as cooperation model. The first implementations will be deployed in confined areas (e.g., small country in which the Operator is active and it is deploying a new infrastructure). This approach will potentially have an impact on existing business model and established relationships between Operators (see for instance Fig. 4.5).
- Traditional Operators see the SDN and the virtualization capabilities as a means to reduce the investments on new infrastructures. In this case, the disruption that SDN is bringing is limited. Also, the fear of the SDN and Virtualization can lower the entrance barriers to such a level that many new Actors can afford the deployment of a network infrastructure. Consequently, they will act in such a way to limit the applicability of SDN and Virtualization.

A byproduct of SDN and Virtualization is a possible change in the Telecommunications equipment market. Indeed, the value moves to software and programmability. New Vendors (e.g., coming from the IT market) can offer appealing propositions displacing the current incumbent Vendors. Established Vendors in the realm of networking as Cisco and Juniper can attempt to mitigate the impact of SDN and Virtualization by exposing and opening up a few APIs in their systems. Traditional telecom Vendors (e.g., Alcatel-Lucent, Nokia Siemens Networks, Ericsson, and others) could try to preserve the value of their solutions by adopting a similar approach. IT Vendors (like HP and IBM) can offer a better proposition because they do not need to preserve existing product lines. In addition, a market of new Vendors can emerge (e.g., VMware with the acquisition of Necira could play an important role in this). In other terms, a disruptive implementation for SDN and Virtualization can lead in a few years to a restructuring of the Telcos' ecosystems introducing new Vendors and transforming existing ones.

From the point of view of a Telco, the combination of SDN and Virtualization poses the problem of Mastering software. It refers to the possibility to directly programming the infrastructure, in order to differentiate from the other competitors. The capability to deal with successful software projects is questionable within Telcos, so they should make a relevant effort at converting their IT department into efficient software factories capable of creating new infrastructural solutions as well as new services and applications on top of them. This transformation requires skills and attitude that not all the Telcos have or are ready to use. This issue strongly influences the possibilities of Telcos to play different roles in the service provision. The two different possibilities actually exist: Telcos as a Platform Provider and the Telco as a Service Provider. In the first case, the Telco will focus on providing services to the final users (and mainly to business ones) while in the latter, the Telco wil focus its attention in creating a programmable and flexible infrastructure that others will use in such a way to govern the network. Both approaches exploit the technologies discussed in this chapter to support higher (compared to today's situation) level of resources and network programmability. Programmability is a requirement for playing the Service and Platform Provider roles.

Another result of this chapter is the emphasis put on the need to create a relation between the complexity of the networked environment and the need to introduce cognitive behaviors in the networks. As said, networks will encompass different functionalities (processing, storage, and sensing besides communications). Networks will become richer in functionalities, but they will be also more complex in terms of allocation and usage of resources and functionalities. In order to cope with this complexity, autonomic and cognitive solutions will be progressively introduced in the network environments so that resources will contribute to the self-organization of the infrastructure. Even if core networks will become flatter and functionally simpler, there will be an increased need to integrate other type of resources and to integrate them even if they pertain to different administrative domains. The infrastructure should be capable of integrating into a virtualized logical network the resources of the Telco as well as the resources of different Enterprises that will dynamically join so as to meet a business goal. Self-organization, federation, and integration of different types of resources will be characterizing features of many future scenarios.

Bibliography

Adeyeye M, Gardner-Stephen P (2011) The Village Telco project: a reliable and practical wireless mesh telephony infrastructure. EURASIP Journal on Wireless Communications and Networking (Springer) 2001(1):1–11

Aoyama T (2009) A new generation network: beyond the internet and NGN. IEEE Commun Mag (IEEE) 47(5):82–87

Arnaud S, Bill JW, Kalali B (2003) Customer-controlled and-managed optical networks. J Lightwave Technol (IEEE) 21(11):2804–2810

Bastos L, Wietgrefe H (2010) Challenges and proposals for WiMAX-based military hybrid communications networks. In: Military communications conference, 2010-MILCOM, San Jose, USA. IEEE, pp 2020–2025

Battiti R, Lo Cigno R, Orava F, Pehrson B (2003) Global growth of open access networks: from warchalking and connection sharing to sustainable business. In: Proceedings of the 1st ACM international workshop on wireless mobile applications and services on WLAN hotspots. New York, NY, USA. ACM, pp 19–28

Berndt H, Minerva R (1995) Definition of service architecture. TINA Consortium, Red Bank, N.J.

Burgess DA, Samra HS et al (2008) The OpenBTS project. Report available at http://openbts.sourceforge.net, http://openBTS.org

Campbell AT, De Meer HG, Kounavis ME, Miki K, Vicente JB, Villela D (1999) A survey of programmable networks. SIGCOMM Comput Commun Rev (ACM) 29(2):7–23

Chaparadza R et al (2009) ETSI industry specification group on autonomic network engineering for the self-managing future internet (ETSI ISG AFI). In: Web information systems engineering-WISE 2009. Springer, Berlin, pp 61–62

CISCO (2012) Visual networking index: global mobile data traffic forecast update, 2011–2016. White Paper, San Jose, Ca, USA, CISCO

Clark DD, Wroclawski J, Sollins KR, Braden R (2005) Tussle in cyberspace: defining tomorrow's internet. IEEE/ACM Trans Netw (TON) (IEEE/ACM) 13(3):462–475

Dunkels A, Alonso J, Voigt T (2003) Making TCP/IP viable for wireless sensor networks. SICS Research Report. Swedish Institute of Computer Science, Stockholm

Fainelli F (2008) The OpenWrt embedded development framework. In: Proceedings of the free and open source software developers European meeting. FOSDEM, Brussles

Follows J, Straeten D (1999) Application driven networking: concepts and architecture for policy-based systems. Red Books. IBM Corporation, White Planes, USA

Fontoura A, Maxim S, Marcus G, Josifovski V (2013) Top-k publish-subscribe for social annotation of news. In: Proceedings of the VLDB Endowment. VLDB, to appear

Fu J, Hao W, Tu M, Ma B, Baldwin J, Bastani FB (2010) Virtual services in cloud computing. In: 6th world congress on services (SERVICES-1). Miami, FL, USA. IEEE, pp 467–472

Gavras Anastasius, Karila Arto, Fdida Serge, May Martin, Potts Martin (2007) Future internet research and experimentation: the FIRE initiative. ACM SIGCOMM Comput Commun Rev (ACM) 37(3):89–92

Geambasu R, Kohno T,. Levy A, Levy HM (2009) Vanish: increasing data privacy with self-destructing data. In: Proceedings of the 18th USENIX security symposium, USENIX, pp 299–316

Gurbani V, Hilt V, Rimac I, Tomsu M, Marocco E (2009) A survey of research on the application-layer traffic optimization problem and the need for layer cooperation. IEEE Commun Mag (IEEE) 47(8):107–112

Jacobson V, Smetters DK, Plass NH, Briggs MF, Stewart P, Thornton JD, Braynard RL (2009) VoCCN: voice-over content-centric networks. In: Proceedings of the 2009 workshop on re-architecting the internet. Rome, Italy. ACM, pp 1–6

Jelasity M, Montresor A, Babaoglu O (2005) Gossip-based aggregation in large dynamic networks. ACM Trans Comput Syst (TOCS) (ACM) 23(3):219–252

Johnson D, Ntlatlapa N, Aichele C (2008) Simple pragmatic approach to mesh routing using BATMAN. Report presented at Wireless Africa, Meraka Institute, Pretoria

Jondral FK (2005) Software-defined radio: basics and evolution to cognitive radio. EURASIP J Wirel Commun Netw 2005(1):275–283

Katti S, Rahul H, Wenjun H, Katabi D, Médard M, Crowcroft J (2008) XORs in the air: practical wireless network coding. IEEE/ACM Trans Netw (TON) (ACM/IEEE) 16(3):497–510

Koponen T et al (2011) Architecting for innovation. ACM SIGCOMM Comput Commun Rev (ACM) 41(3):24–36

Lo Pumo A (2007) Overview of the Netsukuku network. preprint arXiv:0705.0815, pp 1–5

Manzalini A, Minerva R, Moiso C (2010a) Autonomic clouds of components for self-managed service ecosystems. J Telecommun Manag 3(2):164–180

Manzalini A, Minerva R, Moiso C (2010b). Exploiting P2P solutions in telecommunication service delivery platforms. In: Antonopoulos N, Exarchakos G, Li M, Liotta A (eds) Handbook of research on P2P and grid systems for service-oriented computing: models, methodologies and applications. Information Science Reference, Hershey, pp 937–955

Manzalini A, Minerva R, Moiso C (2010c) The inner circle: how to exploit autonomic overlays of virtual resources for creating service ecosystems. 14th international conference on intelligence in next generation networks (ICIN). IEEE, Bordeaux, pp 1–6

Manzalini A, Minerva R, Moiso C (2010d) Towards resource-aware network of networks. In: 5th IEEE international symposium on wireless pervasive computing (ISWPC). IEEE, Modena, pp 221–225

Manzalini A, Moiso C, Minerva R (2011a) Towards 0-touch networks. Technical symposium at ITU telecom world (ITU WT). IEEE, Geneva, pp 69–74

Manzalini A et al (2011b) Self-optimized cognitive network of networks. Comput J (Br Comput Soc) 54(2):189–196

Manzalini A, Nermin B, Corrado M, Roberto M (2012a) Autonomic nature-inspired eco-systems. In: Gavrilova ML, Tan KCJ, Phan C-V (eds) Transactions on computational science, vol 7050. Springer, Berlin, pp 158–191

Manzalini A, Crespi N, Gonçalves V, Minerva R (2012b) Towards Halos Networks ubiquitous networking and computing at the edge. In: 16th international conference on intelligence in next generation networks (ICIN). IEEE, Berlin, pp 65–71

Marinescu D, Kroger R (2007) State of the art in autonomic computing and virtualization. Report. Distributed Systems Lab, Wiesbaden University of Applied Sciences, Wiesbaden

McKeown N, Anderson T, Balakrishnan H, Parulkar G, Peterson L, Rexford J, Shenker S, Turner J (2008) OpenFlow: enabling innovation in campus networks. ACM SIGCOMM Comput Commun Rev (ACM) 38(2):69–74

Moiso C et al (2010) Towards a service ecology for pervasive networked environments. In: Future network and mobile summit. IEEE, Florence, pp 1–8

Peterson L, Wroclawski J (2007) Overview of the GENI architecture. In: GENI design document GDD-06–11, GENI: Global environment for network innovations, Washington, DC, USA, NSF

Pouwelse J, Garbacki P, Epema D, Sips H (2005) The bittorrent p 2p file-sharing system: measurements and analysis. In: Peer-to-peer systems. Springer, Heidelberg, pp 205–216

Rexford J, Dovrolis C (2010) Future internet architecture: clean-slate versus evolutionary research. Commun ACM (ACM) 53(9):36–40

Roberts J (2009) The clean-slate approach to future internet design: a survey of research initiatives. Ann Telecommun 64(5):271–276

Rothenberg CE (2008) The increasing role of the semantic web & ontologies in new generation network architectures. Report online: http://chesteve.wordpress.com/2008/07/01/the-role-of-semantic-web-ontologies-in-newnext-generation-internetworking-architectures/

Rowe D (2009) The mesh potato. Linux J 188, Article no. 5

Slater D, Wu T (2009) Homes with tails-what if you could own your internet connection. White paper. CommLaw Conspec 18:67

Turner J, McKeown N (2008) Can overlay hosting services make ip ossification irrelevant. In: PRESTO: workshop on programmable routers for the extensible services of tomorrow

VMWare (2006) Virtualization overview. White Paper, Palo Alto, Ca, USA

Wan C-Y, Campbell AT, Krishnamurthy L (2002) PSFQ: a reliable transport protocol for wireless sensor networks. In: Proceedings of the 1st ACM international workshop on Wireless sensor networks and applications. Atlanta, GA, USA. ACM, pp 1–11

Weiss A (2006) DD-WRT Tutorial 3: building a wireless bridge. WiFi planet. www.wi-fiplanet.com/tutorials/article.php/3639271. Accessed May, Last accessed 27, 2013

Wilson BJ (2002) Jxta. New Riders—Pearson Publishing, San Francisco

Xie H, Yang YR, Krishnamurthy A, Liu YG, Silberschatz A (2008) P4p: provider portal for applications. ACM SIGCOMM Comput Commun Rev (ACM) 38(4):351–362

Yap K-K et al (2010) Blueprint for introducing innovation into wireless mobile networks. In: Proceedings of the second ACM SIGCOMM workshop on Virtualized infrastructure systems and architectures. New Delhi, India. ACM, pp 25–32

Chapter 5
Emerging Services

Information technologies, such as cloud, IoT, virtualization and smart environment, shape new business. Traditional services (e.g., RCS) fail to get traction, compared with new cloud opportunities. While Voice migration to webRTC offers low payback, there is more scope in acting as storage providers or data broker. Although "cloud+network" is challenged by major web cloud providers, Telcos can still offer federated clouds solutions to large enterprises and vertical markets.

Cloud information must be respectful of data ownership and privacy, but users can be empowered to "store-and-share." Personal data is now the new "currency" by which users pay for services (bank-of-users). Telcos can be reliable brokers, which aggregate and anonymize the data before selling it. The same applies to "Smart Environments" that combine data from different sources, including sensors and mobile phones, with different ambience, awareness, and data merging.

Mobile devices are increasingly used as sensors, with users sharing collected data, as in crowdsensing. Devices that communicate via IoT can become Cloud "virtual objects," with added value of data integration that deliver sought-after applications, thanks to virtualization. Terminal virtualization can reflect physical objects that are also linked to identity management and personal or shared aggregated data, enabling development of rich applications.

5.1 Introduction

The Technological evolution described in previous chapters will lead to new opportunities in terms of services. These opportunities are related to new classes of services that could not be proposed in the past due to technological limitations (e.g., augmented and virtual reality), to current services like the cloud, and to existing classes of services (like communications services that could be totally reshaped by the new technology). This chapter does not aim at covering in an exhaustive manner the possibilities,but it tries to evaluate a few interesting classes of services that

R. Minerva and N. Crespi, *Networks and New Services: A Complete Story*,
Internet of Things, DOI 10.1007/978-3-319-33995-5_5

could have a value from a Telco perspective (see for instance Sect. 6.5 in which some lines of action for the operators are presented), the impact that new technologies could have on existing services, the possible evolution of existing services and the introduction of new classes of services. One feature of these classes of services is that they have a potential large market that could help the Telco in finding alternative markets to its traditional communication service market. Still it should be clear that there are also other possibilities like multimedia services and evolution of IPTV, or Virtual and Augmented reality applications, or even e-government-related applications. All of them are large potential market and they deserve a careful attention. The classes of services discussed in this chapter are those under close evaluation of a number of operators (e.g., Telecom Italia and Telefonica together with MIT for personal data), the evolution of machine to machine toward Internet of Things undergoing within numerous operators. The cloud represent a specific example: many operators need to correctly position their offering in this area in order to understand whether the network and ICT assets of a Telco can be exploited. Smart Environments are instead an initiative undergoing within Telecom Italia whose goal is to understand how edge environments (and direct to direct communication) could be integrated within the Telco operator with an open and possibly user-centric approach.

The objectives of this chapter could be summarized as follows:

- To present some new classes of services that can provide new opportunities to several actors (the analysis is focused on the Telcos)
- To highlight the way, these services can be implemented and the merit that interaction and control paradigms for supporting them could have (and vice versa the difficulties that can be met in trying to implement services choosing a paradigm "a priori").
- To identify if possible new user-centric approaches and their viability in the provision of new services. These aspects are particularly relevant in the personal data management that is considered by many actors as one new important source of revenue and a sort of test-bed to establish new forms of relationships with users.

The chapter focuses on the opportunities and the challenges that cloud computing is posing to Telcos. This is a main trend, in fact many Telcos want to pursue this business in order to increase their value proposition to customers. The chapter will indeed analyze what conditions and what approaches are indicated in order to take advantage from these technologies. In addition some considerations on the Cloud Ecosystems are presented.

In this section, a few Service and Application Scenarios are presented in order to frame the possible technological evolution in viable deployments or use cases. This section is not necessarily advocating the merits of a service or class from the specific point of view of a stakeholder (e.g., a Web Company, and Operator), it offers mainly a view on what could be done at the technological level. Whenever it is possible, the chapter emphasizes a user-centered proposition, i.e., a proposal that puts the users and their rights at the center more than the specific economic return of a stakeholder. However, some examples could have a Telco-oriented approach as a means to identify some viable services and applications that can bring value to the customers and can help the operators to find different roles and business models in the ICT markets. Service scenarios neither aim at being exhaustive nor complete, they just cover a set of areas that promise interesting developments and give an idea of how solutions could be forced by specific business models or approaches.

Telcos are and will be interested in communications services, so they will pursue the goal of creating new communications capabilities. There are several attempts going on to exploit existing means in order to increase the appeal and the number of functionalities offered by communication services. The term Rich Communication Service refers to the possibility to aggregate around the "data transport" capability a set of appealing functions, like VoIP, Presence, Instant Messaging, and the like. This is another example of the general category of Unified Communications. The Telcos are striving to revive this market and they are trying to exploit existing architectures (like IMS with the Rich Communications Suite) or new ones (like the rising WebRTC solutions).

Another recurrent intention is to enter into the loud computing market offering several XaaS possibilities. One of the assets to capitalize, in this case, seems to be the possibility to locally serve and support the customers. In addition, Telcos are interested in offering federated cloud functionalities in order to integrate customer solutions in several different countries. There is a need to understand whether these proposals are interesting for the user and how they can be differentiated and improved with respect to offers made by WebCos. In addition, there is the issue related to the technological gap between WebCos and the other possible players of the cloud. In this case, the completion has to be played at the technological and business level in a very aggressive fashion.

On the new services side, there are some opportunities that the Telco may undertake. The Bank of User Data is one of them: personal data will be collected according to policies and mechanisms established by the user and the ownership of data remains of the user. The Telco can play the role of storage provider and possibly the one of broker of data if the user enables this possibility. Plenty of services could be created and offered on personal data infrastructures. Another example is the Smart Environments, i.e., intelligent systems that can adapt their resources in order to support the communication, processing, and storage needs of several users in a dynamic way. Finally, another opportunity is presented: the Internet with Things, i.e., the capability to create a networked environment in which any physical resource can be virtualized in the clouds creating a sort of relationship between the atoms (of the physical resource) and the bits (of the virtualized one). Actions on the physical or virtual world have an impact also in the "other world." This approach is very different from the Machine to Machine (M2M), one. Today Telcos are using M2M platform to earn connectivity from distributed devices and from management services (dynamic activation and deactivation of SIM cards). The level of functionalities provided is very limited and with the increase in pervasiveness of other solutions, this opportunity can vanish if it is not supported by a richer set of functionalities. Internet with Things seems to be a vast domain into which to offer new compelling services.

Determining the opportunities of providing new services is a means to understand if there are significant business returns and a means to understand how much to invest into service architectures. One of the issues limiting investments in horizontal platforms for instance is the need to justify the platform costs with respect to vertical solutions on a service-by-service basis. Horizontal platforms make sense if and when the number of services that can be potentially offered is huge and these services are easily supported on a delivery platform. These classes of services are promising also from this point of view.

5.2 New Communication Services

Since the inception of Voice over IP solutions (e.g., Voicetec in the ninety), the world of telecommunications has seen these technologies with suspicion and fear. It was clear that they were capable of changing the way the major services of Telecommunications networks were provided to customers and were paid for.

Even new technologies are often modified and conducted to a specific model: the network intelligence. This is a constant attitude of many operators that try to smooth the disruption of new technologies with respect to communication services. There is a long tradition in this approach.

The major attempts to deal with Voice over IP (VoIP) technologies were to frame them in the usual architectural vision as represented in Fig. 5.1 (Vemuri 2000).

Services are value-added functions provided by specialized nodes (in this case the Intelligent Network Controllers) and service functionalities are specified and provided according to a standardized finite-state machine for representing call progress.

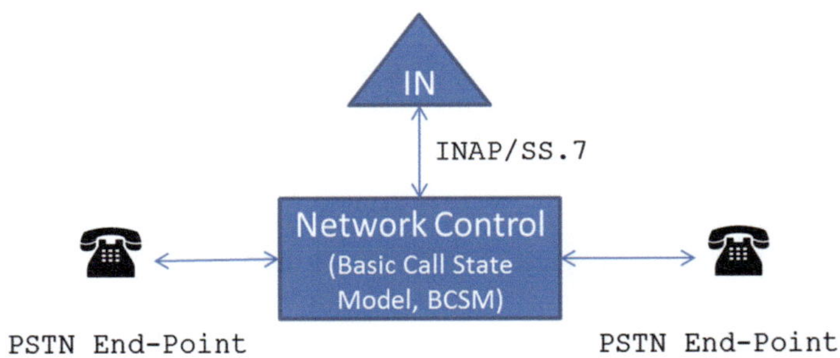

Fig. 5.1 The network intelligence paradigm

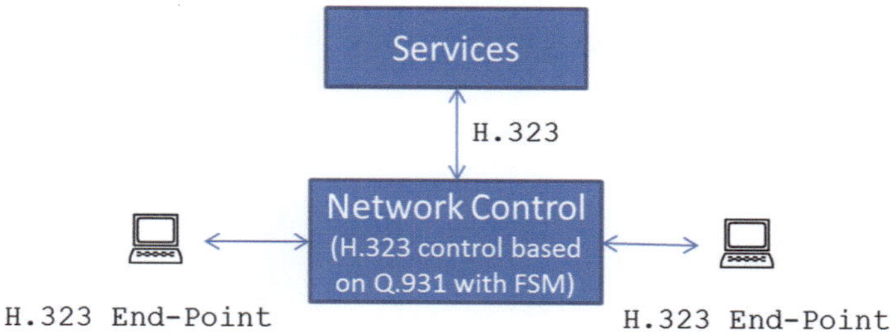

Fig. 5.2 The H.323 control architecture

For instance, the H.323 protocol stack was designed to be compatible and to be used as a mere evolution of the voice architecture. Actually, H.323 protocol stack is an adaptation of the telephony signaling protocols over an IP signaling infrastructure. The paradigm behind it is essentially the same traditional one as represented in Fig. 5.2.

Even the Session Initiation Protocol (SIP) (Rosenberg et al. 2002), has been modified and changed within 3GPP in order to better fit into the paradigm of the network intelligence. Actually, H.323 (Thom 1996) and SIP (Schulzrinne and Rosenberg 1998) have been considered as the new signaling system, i.e., a means for end points to ask the network how to orchestrate resources in order to put in place a "call" or in more modern terms a "session." Under this respect, call and session are synonymous. They refer to the fact that endpoints and network functions/resources are correlated and associated with the goal to create a suitable communication environment for the time requested by the participants. H.323 was already a stateful protocol with its own call model. SIP can be stateful and stateless, i.e., its servers could keep track of the interactions with end points and behave according to a complex session state machine. This approach has been pursued in the 3GPP specifications leading to some changes in the SIP definition for adapting this protocol to the "network intelligence" paradigm, i.e., services are to be provided within the context of a session, they are related to the allocation of some network resources, service logic is triggered by the session control function when it has the need to do so. The result of this redefinition of the SIP protocol is showed in Fig. 5.3.

This attitude and philosophy has left room for others to innovate and adopt more distributed approaches. Skype is paradigmatic from this point of view and it shows how P2P mechanisms can support large-scale communication systems. It is based on a hybrid and structured peer-to-peer approach in which the end-to-end communication takes place between the peer nodes, while a part of the signaling and the organization for the overlay network relays on the functions offered by "super-nodes" and the identification, authorization, and billing functionalities are centralized into a few servers directly under the control of Skype (Fig. 5.4). The architectural choices are quite smart; the valuable and important functions from a business perspective (Identity, authorization, and billing) are kept in a centralized and well-controlled infrastructure. Other control functions (like addressing, routing,

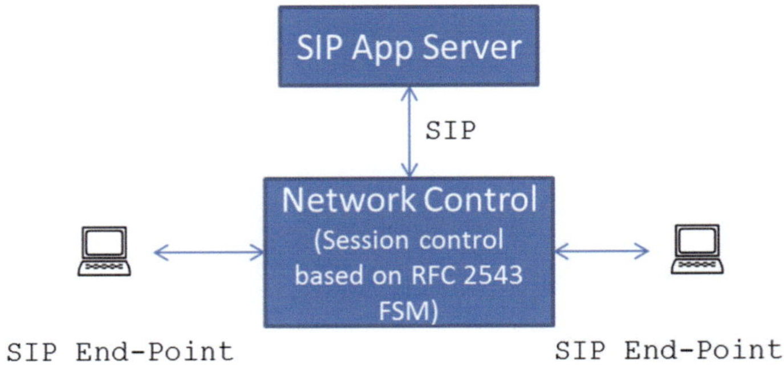

Fig. 5.3 The network intelligence approach to SIP

Fig. 5.4 The skype architecture

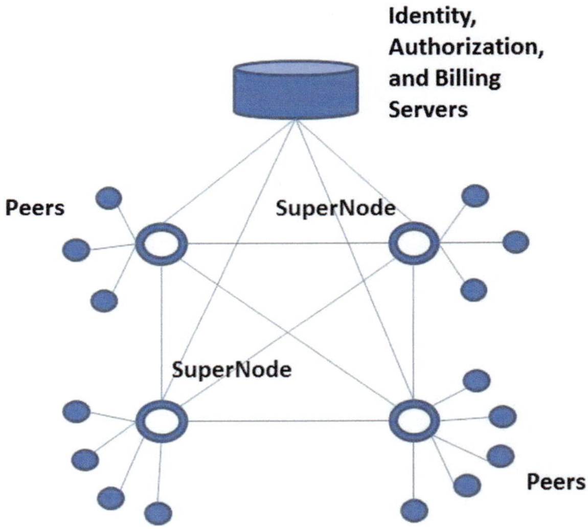

grouping of peers, their organization, and "keepalive") are demanded to a set of supernodes (i.e., trusted node collocated in "friendly" locations). The peers have the control of local linkage to supernodes and the end-to-end communication with other peers.

The interesting fact is that one of the fathers of the SIP protocol had to go through a reverse engineering study in order to determine the structure of Skype (Baset and Schulzrinne 2006). The work was aiming at determining the reasons of the success of the P2P solution. Contextually in the same period, the SIP community was trying to propose the usage of SIP as a viable protocol to create P2P systems (Singh and Henning 2005). The Skype implementation is based on: the consolidated Joltid[1]

[1]http://joltid.com/. Last accessed May 30th 2013.

platform used in Kazaa[2] for supporting a large-scale P2P system, the economic implementation, the quality of codec used. All these features were missing at the time of the SIP protocol (that once again was used in a very traditional manner). Currently, Skype is dominating the VoIP market and has reached a rewarding position as a collector of international traffic (around 34 % of international traffic with a record of over 50 million concurrent users according to (Skype, Wikipedia 2013), and it has added over time many interesting features, like presence, video call, video conferencing, and desktop sharing.

Operators are still striving in determining how to regain a grip on users. There is an effort in positioning the RCS (Henry et al. 2009; Lin and Arias 2011), as designed and put together by the GSM Association (GSM Association 2009). The attempt is to provide an interoperable solution between several operators able to support presence, instant messaging, synchronization with address book, and other value-added functions by means of APIs and using the IMS infrastructure. The solution targets smart phone with the intent to provide services to advanced users. The issue behind RCS are several: the architecture is not proven itself so flexible and the issues in interoperability between different IMS systems are causing interoperability problems (that major Web Companies do not have); services are not differentiating from those offered already by WebCos and, in addition, they suffer from a bit of invasiveness (e.g., the presence feature was originally offered in such a way to violate the users privacy); the service is not based on a clear definition of a community (like Facebook), and hence, they are very similar to the "old way" of making phone calls. Simply put: from a user-centric view, the RCS does not offer any novelty neither from the service functionalities point of view nor from the business perspective (it is using the existing billing models) and nor from the platform one (the IMS and service exposure combination). In addition, also the phone makers are "tepid" in the initiative.

In more recent time, there is interest in the Telcos community for the applicability of the WebRTC (Loreto and Romano 2012) solutions for supporting the real-time communication between browsers. WebRTC is under definition by the World Wide Web Consortium (W3C). The specification is based on an API and supporting communication protocols enabling "browser to browser" applications for voice call, video chat, and P2P file sharing without plugins. In a way, WebRTC expands the capability of the WebSockets, i.e., mechanisms that are supported in HTML5 to allow the bidirectional communication between a browser and a WebServer avoiding the continuous polling. The envisaged use of WebRTC is by means of a Server that sits in between the peers for helping in establishing the communication, once this goal has been achieved, the peers (the browsers) can directly communicate for exchanging video, audio, and data. Figure 5.5 shows the described mechanisms. It should be noted that the definition of WebRTC supports also the possibility to have a direct signaling between the web browsers (a real P2P solution), however, many actors like web companies and operators will find more convenient to have a

[2]http://en.wikipedia.org/wiki/Kazaa. Last accessed May 30th 2013.

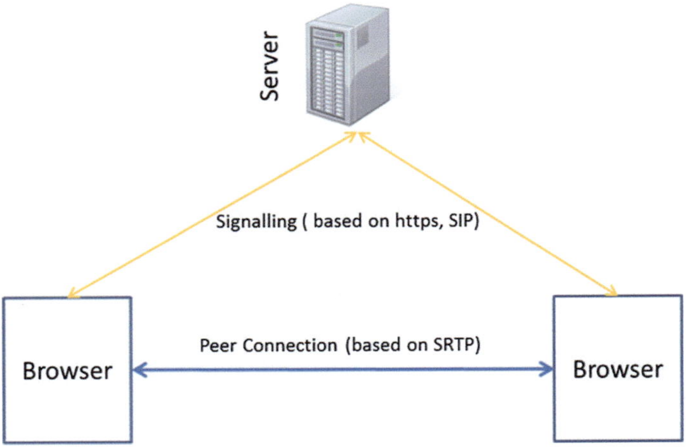

Fig. 5.5 The WebRTC architecture

brokering function between different peers. Obviously, this is a replication of existing models reproduced in order to have a controlling position over users.

From the programmability side, the WebRTC solution is interesting, in fact if offers an API integrated within the browser so that applications can use it to control the setting up of the communication between the peers (with the support of a server) and the transport of data, video, and audio between peers. Figure 5.6 shows the software architecture within a browser.

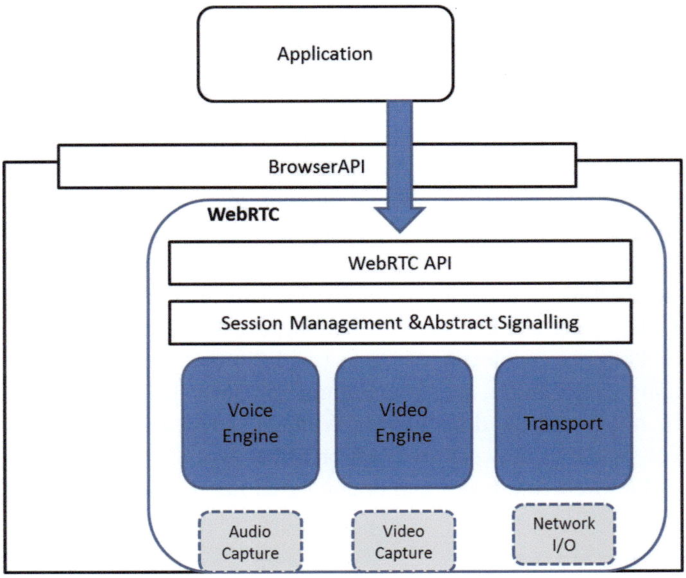

Fig. 5.6 The WebRTC software architecture within a browser

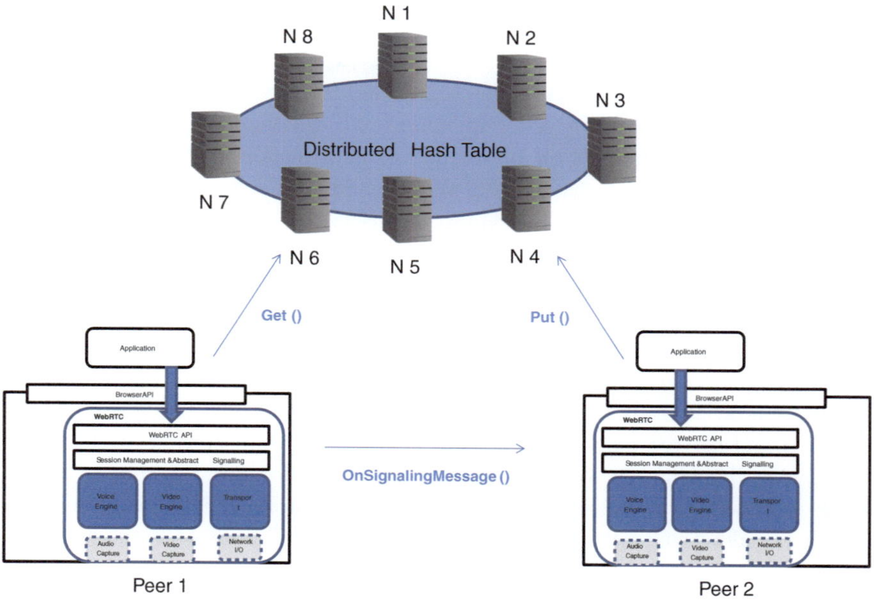

Fig. 5.7 Using WebRTC in a DHT context

It is pretty obvious to think about an extension of this architecture in order to make it highly distributed and consistent with a structured P2P network as Chord. In this case, a distributed hash table (DHT) can be used to find the address of a specific user and then to connect the end points directly without the need of having intermediation by a signaling server (Werner 2013).

Figure 5.7 shows the possibility to use a DHT [like openDHT and its interfaces (Rhea et al. 2005)] for storing and retrieving information needed to connect to a specific peer.

In this example, `Peer 2` stores its information in the DHT (Put() operation). `Peer 1` retrieves the information from the DHT (Get() operation) and then sends an invitation to the other peer for establishing the connection. This example, very far from being complete, just shows a possible trend that many web programmers could endeavor aiming at a more user-centric approach. A few considerations emerge from the analysis of the approach of Telcos to new communications services:

- P2P paradigm is capable to compete with the more traditional Network intelligence one as shown by Skype
- WebRTC can be easily used in a P2P fashion, leaving out the operators from browser-to-browser voice communications
- The application of the network intelligence approach do not give advantages to user in terms of service functionalities, it just helps the Telco to keep a grip on customers.

5.3 From Cloud to Everything as a Services (XaaS)

This section analyses the cloud technologies and its service opportunities mainly from a Telco perspective in order to understand whether the operators have a chance to introduce innovation or disruption in a market that (at least for the residential customers) is in the hands of WebCos. Many Telcos see the cloud and its evolution toward the approach Everything as a Service (XaaS) as an opportunity to enter into adjacent markets, especially, when the cloud is associated with the network asset (an advertisement from Telecom Italia said: "the Cloud with the network inside"). This approach is quiet controversial because (as seen in Sect. 3.3) leading WebCos have a considerable technological advantage over competition and there are doubts that leveraging the network asset will give to Telcos an advantage. In this section, an analysis of what Telcos could really do and achieve in this technological and service realm is presented. One real point to consider is to clearly show what benefits an operator can offer to customers by means of the "cloud + network" proposal. In other terms, the technological proposition has to provide an added value with respect to the advanced offerings of WebCos.

5.3.1 Cloud Computing Services from a Telco Perspective

As seen, the cloud computing technological and market scenarios are largely dominated by Web and Information Technology companies. The technological pace is determined by needs and solutions stemming from the web companies that were able to create walled gardens with proprietary technologies. Each major web player is also able to directly and autonomously develop and master its own specific solution. From this perspective, the technical gap between the Web and the Telecom industries is striking and probably insurmountable. In addition, major web companies have a significant footprint in the provision of services to residential users and their services are deperimeterized (i.e., they can be accessed independently from an owned network, see Sect. 6.5).

Competition under these circumstances is hard, especially, if Telcos are continuously playing the role of "intelligent buyers" and do deliver services on platforms developed by IT or Telecommunications companies.

In order to improve this situation, Telcos should change the rules of the game at the technological and at the business level.

In the following sections, a divergent perspective on cloud computing for Telcos is presented and discussed. It is based on the assumptions that market and customer differentiation is a premium, there is not only a "pay per use" business model behind cloud computing, and residential and business market segments have different needs and expectations. From a technical perspective, the integration of connectivity within private and federated clouds could be a key element for bringing cloud solutions to enterprises, which network programmability is a means to support

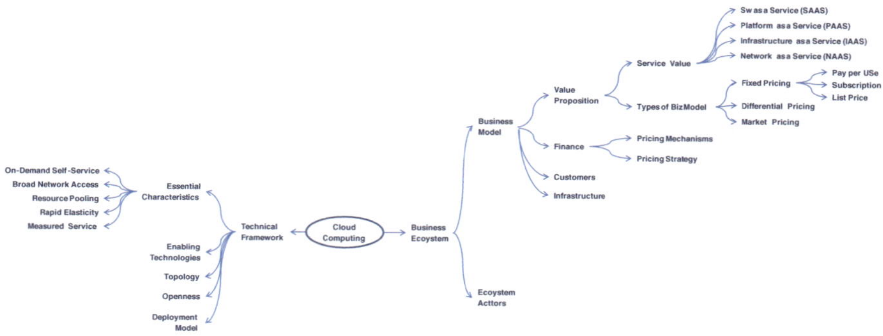

Fig. 5.8 New viewpoints on the cloud taxonomy

enterprise requirements and a pass through for delivering better distributed services that can support consistency of data when customers require such a feature.

Approaching the cloud computing in this different way means also to have another view on the taxonomy of NIST (Liu et al. 2011). In fact new dimensions and aspects of the cloud computing proposition could be introduced. Figure 5.8 illustrate a few new viewpoints that an operator must consider while entering into the cloud competition.

From a business perspective, some new considerations for the cloud market are related to the value proposition to associate to the service models, some considerations related to pricing mechanisms and even more important to pricing strategies. A better definition of target customers could help in tailoring solutions that fit the expectations of the users. The business infrastructure identifies the mechanisms and the processes that a Telco can leverage to pursue a cloud-related offering. And finally, the actors of the ecosystems are those stakeholders that have relationship with the Telco and can help or contribute or have to be involved in order to make a viable business.

On the technical side, new dimensions are related to enabling technologies to be used for a viable implementation of a cloud infrastructure able to support the business objectives. Topology deals with the organization and the kind of infrastructure to be controlled. Openness point to a major feature of the platform: the ability of the proposed platform to be extensible and flexible in such a way to extend and improve the functionalities and services provided over time. The deployment model extends a bit the one proposed by NIST and tries to figure out some viable and meaningful deployment from the Telco perspective.

5.3.2 *On the Cloud Business Ecosystem*

Two aspects of the Business Ecosystem will be briefly sketched in this section: the aggregated actors, and the business models and opportunities reasonably pursuable

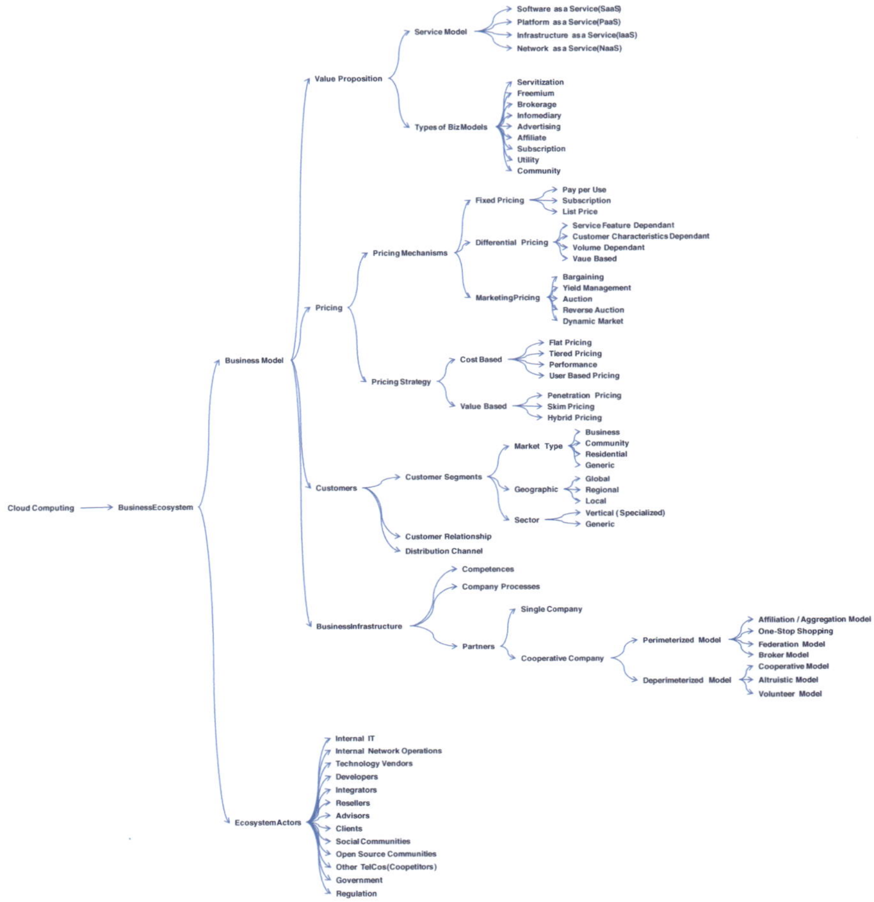

Fig. 5.9 Taxonomy of cloud computing from a business and Telco perspective

by a Telco. They strongly characterize the possibilities and the approaches that a Telco can attempt.

The Ecosystems Actors

The ecosystem of a cloud offering (from a Telco perspective) is quite complex because Telcos do need to have a direct link with customers and because the construction of the Telco cloud requires a lot of links with other stakeholders. In addition, Telcos are not relying on a "make," but on a "buy" approach and then the construction phase of a cloud solution is made also of relationships and integration with other entities. Telcos have to seek cooperation of a large number of other stakeholders in order to put in place a cloud offering. Figure 5.9 (bottom links) depicts a possible set of stakeholders for a cloud platform.

Central to the approach is a clear and valuable relation with the users. It is mandatory to be able to have a different approach to customers compared to the one established by web companies: in this case, there is a need to have a direct communication with customers in order to support them, to integrate their systems and to fulfill their requirements by means of a day-by-day cooperation. Clients in this case are also Enterprises that seek a greater integration of their private platforms into a cloud. Communities (in a large sense) are also important in order to grasp requirements, to promote the solution and the functionalities to a large audience, and for extending and tuning the offered capabilities (a sort of beta test). From a development point of view, the internal IT and Network organizations of a Telco have to cooperate in order to design and agree the best specification for a cloud platform, they should also cooperate in order to define a target list of services and the conditions for integrating the platform and its systems into the Telcos processes and workflow. Other actors involved in the definition, design, and implementation of the platform are technologies vendors, developers, and integrators. Here the goal is to avoid as much as possible a lock in situation in which the Telco is forced to follow the design and product evolution decisions of specific vendors. In such a competitive market (in which the Telco is not the primary choice for many customers), flexibility and readiness to modify and extend the capabilities is of paramount importance. Advisors and consultancies agents should cooperate in these phases in order to advice on trends and best practices of the industry. From a more commercial point of view, Resellers and even other Telcos can be useful to enlarge the potential market of the cloud platform in order to exceed the rigid boundaries determined by the need to deploy networks in specific geographic areas. Government and Regulation Authorities have a role in governing the possibilities and the limits of Telcos in this market. Governments can also be seen as potential customers for the usage of cloud solutions in many situations.

The current business model of many WebCo is the one of a walled garden. Telcos have to nurture new ecosystems in the field of cloud computing allowing a more open model for application development and for the integration within customers systems. Interoperability between different systems and environments and new interfaces are important in order to catalyze new developments that are portable over different cloud computing platforms. The expected contribution from enterprises and developers is the creation of a federated environment that is open and can exploit and leverage the contribution of each stakeholder.

The Business Model

The Business Model definition should encompass at least a convincing value proposition, i.e., a clear definition of the value chain and the perceived benefits for customers and the other stakeholders. This is a combination of the service model (XaaS) and the types of Business Model applied to the service model. As described in (Armbrust et al. 2010), there are several business models for cloud computing and the Internet. For a complete list, refer to (Rappa 2004). The Telco proposition should be aligned to the real possibilities that an Operator has in the market. Some types of business models are out of scope (such as Manufacturer, or others) while

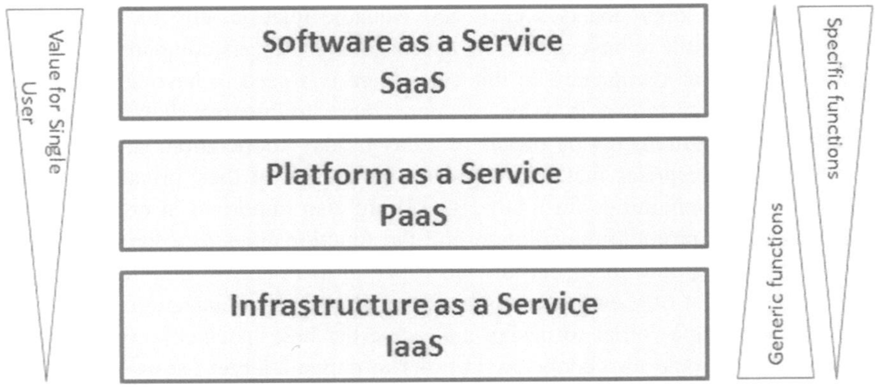

Fig. 5.10 The value of cloud layers and functions as perceived by users

the Broker one seems to fit well in the tradition and skills of Telcos. Another viable option is the possibility to help customers to better enter in the realm of servitization by means of a cloud solution, i.e., the Telco, the customer and the users can establish a Business to Business to Customer (B2B2C), relationship in which the cloud platform enables the customer to move from the selling of products into the selling of product-related services. The capillary presence of Telcos in the territory and the support of customer care department can make this even more appealing and possible.

Pricing is another important aspect of the construction of a sustainable business model. Different pricing schemes can be applied to cloud services. They can have a fixed price structure in which the user pays for resources usage, or a subscription fee. However, cloud computing can be charged also according to the perceived value of the customer of the services provided by the Telco. Figure 5.10 illustrates the value that customers give to the different service models.

A hybrid pricing model could be offered to customers: basic and generic functionalities of the Infrastructure and platform could have a fixed price or a differential one (depending on quantities and volumes), while specific and tailor services could be feature dependent. Other interesting options are related to the possibility to dynamically adjust the price by means of auctions or bargaining with customers for resources and features made dynamically available. For instance, Google is using a complex auction mechanism in bidding for personalized advertisement; a similar approach could be adopted for allocation of valuable resources of the cloud. Other pricing strategies should be carefully analyzed in order to align the pricing schemas to the cost structure of the cloud solution that the Telco is building.

Another relevant issue is determining the target customers. Web companies have supremacy in the customer market and a strong grip on Small and Medium Enterprise (SME) market, but sometimes they lack the same hold on larger businesses. One possible step for the Telco is to address mainly the business market by leveraging its local customer management and channel distribution capabilities. In addition, Telcos could differentiate services and features of the cloud platform in terms of vertical markets, i.e., the cloud offering could be instrumental for many businesses for better cover and exploit specific markets (such as Public Administration, e-health, smart cities, and the like). Another major point is the possibility to focus on a national market or to have an international footprint. In this latter case, the Telco should create a network of relationships and allies in those markets in which a direct presence is not possible. In addition, the Telco could take the opportunity to leverage different deployment models in order to create a global cloud solution. As a rule of thumb, a Telco should adopt and promote the federation of cloud solutions in order to create a global coverage of the market, and at the same time, it should act locally by promoting hybrid solutions in local markets. The combination of a federated and a hybrid approaches allows to provide customers a large portfolio of standardized and generic services. These services could be locally tailored to the needs of the specific market or even customers. This flexibility and elasticity should be supported at the technical infrastructure level by a high degree of programmability and composition of services.

Eventually, a Telco should take care of its Business infrastructure, i.e., the combination of skill/competences, processes, and company attitude in doing business. Processes and IT skills are quite important for a successful deployment of cloud infrastructures, however, they should also be supported by the right capability of doing business in a cooperative way (e.g., by involving partners or with a "doing yourself" approach). Another dimension of the Business Infrastructure is the willingness to pursue an open or a closed model for the cloud platform. A closed way of operating in the market naturally excludes deperimeterized coverage because the company is reluctant to operate in within a dynamic scenario of short term or opportunistic relationships and alliances. In this case, a perimeterized model (and market) is more appropriated. The previous Fig. 5.9 summarizes some of the aspects related to the Business model dimension.

5.3.3 *On the Cloud Technical Framework*

The NIST technical framework (Liu et al. 2011) under which cloud solutions can be designed and implemented is extended in this section. The reasons for this broadening lay are need to better leverage from a Telco perspective the network assets and to promote them to the general attention. This leverage is not pursued with a traditional perspective (i.e., the network has value and it provides Quality of Service related features); instead the networking capabilities are framed within a highly distributed environment compatible with the end-to-end principle of the Internet.

One of the beliefs is that networking aspects will be more considered in the future of cloud computing and will not be treated as minor issues in the provision of cloud solutions. If Nielsen's Law holds true (Nielsen 1998) (i.e., a high-end user's connection speed grows by 50 % per year[3]) then the users will be limited by the network capabilities. This means that the growth in bandwidth will lag behind the growth in processing power. From a cloud computing perspective, this law could have interesting consequences: the commoditization effect on processing will be faster than on bandwidth (i.e., 18 months for doubling the processor power vs. 21 months to double the available bandwidth); bandwidth could maintain a premium value over computing. If this law holds valid than a sort of **paradox** could emerge: cloud computing providers started by providing a value-added service, but they will end up providing a commodity service, while Telcos are starting by providing a commodity service and will end up offering a value-added connectivity service needed to the whole cloud ecosystem. Scarcity of bandwidth could be an important factor for the optimization of network resources.

The enabling technologies of cloud computing cover a broad spectrum. They range from virtualization up to data management. Virtualization has been applied so far mainly to processing and storage. New emerging technologies are bringing virtualization benefits also to other kind of resources: networking resources have been covered discussing the advancements of projects like OpenFlow in Sect. 4.4. Smart objects and sensors will be virtualized as well. This will lead to a decoupling of local proprietary sensor solutions from the virtual representation in the cloud of virtual and smart objects. The combination of virtualized communication, processing, and storage capabilities coupled with smart objects within the cloud will make possible new kinds of applications in the fields of Internet of Things and ambient intelligence. Any real object could be in the near future virtually represented by means of a clone in the cloud. The relationship between a real and a virtualized object creates a sort of continuum that allows to interact, manipulate, and govern real objects that can be augmented in terms of features and intelligence. The number of distributed smart objects will increase over time and it will soon become difficult to control and manage them by means of human intervention. These objects will progressively expose intelligent behavior and the ability to self-organize in complex situations becoming autonomics. Autonomics capabilities and ubiquitous communication will make these objects pervasive. Pervasiveness will determine the possibility to be involved and actually support and control large parts of production life cycles, or processes in the home or in the enterprises. Their strong relation with cloud computing will bring an increase in the perceived value of cloud-based applications and services. Objects will need to communicate each other in order to adapt to the execution context and to the desiderata of end users. The topology of cloud infrastructures will change because pervasiveness, heterogeneity of intelligent entities, and objects governed by cloud services, dynamicity, and elasticity of service environments will require the ability to integrate different resources into autonomic and intelligent systems. They will be arranged in a distributed fashion, but for specific applications, there will be the need to centralize resources and architectures (e.g., government related applications). In addition,

[3]See http://en.wikipedia.org/wiki/Nielsen%27s_law#Contributions. Last accessed May 30th 2013.

(virtualized) resources are to be programmed and controlled; extensibility and programmability of resources will be a common requirement. In environments operating closely with the final customers, there will be an increasing need for trust also in the software development. Open-source implementations could find a further boost because they are controlled and extended by large communities of programmers that continuously check for bugs and malicious developments. Deployment models will greatly vary from closed environment (e.g., private clouds) to federated solutions or even “clouds of clouds” (i.e., interclouds) in which capabilities and resources of heterogeneous and different infrastructures will be negotiated and then integrated in a seamless platform. This evolution will emphasize the need for interoperability between different clouds, the value of connectivity for creating dynamic links between resources and consequently will require trusted and regulated stakeholders. Telcos will have a chance to play a relevant role in this context by leveraging assets, such as connectivity, identity management, and a regulated behavior that is respectful of users privacy.

Cloud computing will also extend its essential capabilities by offering choices between consistency and availability of data. Programmers and developers of services will be able to choose between different solutions for guaranteeing transactions and consistency to financial services or high availability and real-time speed data stream management. Terminals will be more and more integrated in the cloud, initially as simple clients but progressively they will become active nodes in the provisioning of services.

Figure 5.11 is providing a view on the technical framework of the upcoming cloud computing.

5.3.4 Examples of Telco Services in a Cloud Context

Telcos have to use the cloud approach for providing services mainly to enterprises and business customers. This choice is dictated by the following reasons: business customers are more interested in creating a long-lasting relationship with the provider, this link can also leverage customer relationship management systems; the existing billing relationship for connectivity services can be exploited in order to promote a sort of one stop shop approach for many services; the Telco can leverage the local footprint and to provide integration capabilities at a global level, for certain enterprises these capabilities can make a difference.

The Telco’s approach to the cloud has to promote and leverage the interoperability and integration of different complex IT systems into a federated infrastructure with a rich service portfolio. However, this large infrastructure should be flexible enough to accommodate for private systems. Enterprises should be able to decide which processing, storage, communication, and sensing capabilities to keep in-house and which ones to externalize.

In the following section, examples of cloud services (in general according to the Platform as a Service, PaaS, or Software as a Service (SaaS) models) based on

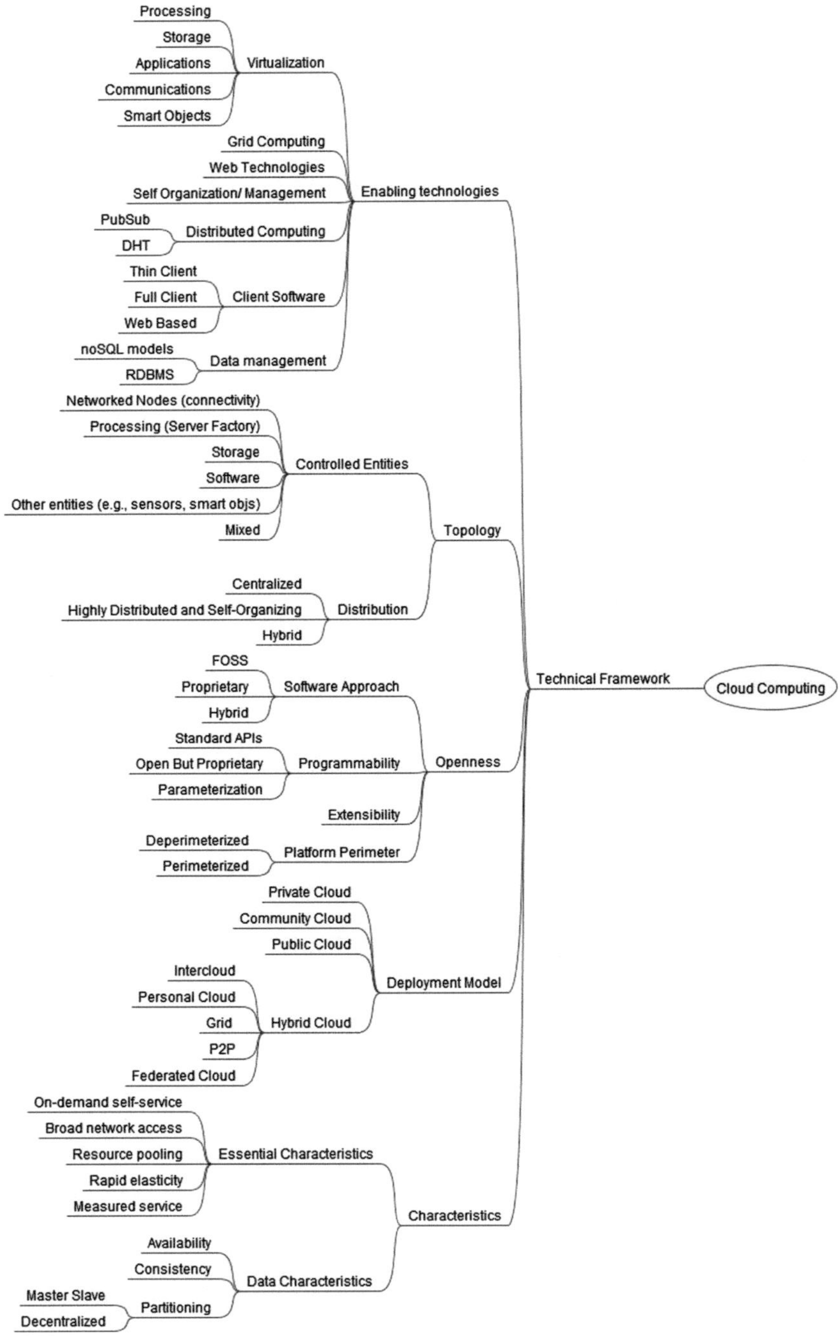

Fig. 5.11 A taxonomy of technical aspects of cloud computing

these foundations are given. A Business to Business to Customer approach is usually pursued in order to leverage the assumed capabilities of a Telco's cloud platform.

Intermediary in the Internet with Things

The Internet of Things is an environment in which some intermediary functionality can have value. As proposed in (Bohli et al. 2009), Telcos play the role of intermediary between small (wireless) sensor network providers and final customers. The operators could even promote the wide adoption of (wireless) sensor networks by subsidizing the small sensor providers. The goal is to collect a set of meaningful data that can be exploited by determining macro trends within a community or a specific location.

In particular, the cloud platform could be extended for offering functions aiming at a "Data Cloud":

- to collect data from sensors networks geographically dispersed and to integrate them with other data sets (e.g., from the Public Administration in an open data fashion, from users' devices, and automotive embedded devices);
- to store large data sets to be dynamically updated and to analyze (data mining) and aggregate data into information;
- to provide message-based engines and to derive information from complex events management;
- to distribute to service providers' applications relevant and updated data with the requested aggregation level and according to the agreed dispatching mechanism (e.g., pub-sub, continuous query, big tables, …).

This service could be characterized by:

- frequent interaction with aggregation systems of distributed sensor networks;
- elasticity in processing and storage allocation depending on data and processing to be dealt with;
- interaction with centralized and traditional applications;
- Data cloud capabilities offered as PaaS and SaaS.

Figure 5.12 describes the service.

Massively Multiplayer Online Games (MMOG)

MMOG Providers at a global level could request cloud computing services aiming at:

- Optimizing the processing and storage load balancing;
- Optimizing the network features in order to provide a better service to their clients, e.g., optimization of latencies for the users, dynamic increase in bandwidth allocation (e.g., to download large files), etc.

The cloud computing provider, in addition, could provide specific functions for the MMOG Provider such as:

Fig. 5.12 The intermediary service in an internet of things context

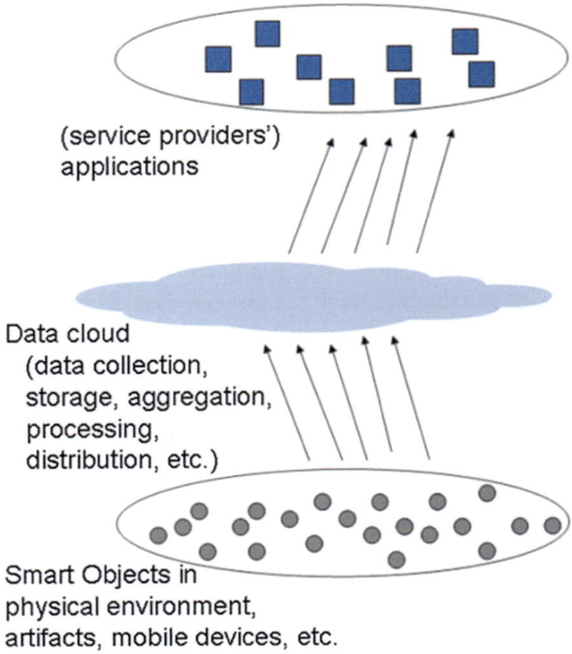

- Load prediction service, in order to evaluate the servers load based on a prediction of distribution of entities and players;
- Resource allocation service: to provide local servers and rebalance the processing load due to an increasing number of players.

The service should aim (maybe in a Silk fashion) at rebalancing the processing and storage load in order to fulfill real-time constraints and real-time capabilities as negotiated with the users.

Features of this service are:

- support of QoS parameters in the interactions with user terminals;
- load balancing and optimization capabilities.

Similar features could also be used to better support multimedia service provision within a specialized cloud.

They could be packaged in a specific PaaS offering for MMOG or generic Multimedia Providers.

Virtual Terminals

Physical devices could be virtualized in the cloud and augmented with additional features or capabilities (e.g., more processing power or more storage). The physical device could use its virtual image for the execution of background processing; for the migration (teleporting) of tasks that do require too many physical device resources (Chun et al. 2011); migration of tasks that requires more resources than those available in the physical device or that have nonfunctional requirements (e.g., performance, security, reliability, parallel execution) that the physical terminal cannot satisfy; delegation of tasks to be executed when the physical device is not connected to a network; extension of storage for keeping all the events forwarded or generated by the terminal. Feasible scenarios are related to the usage of this service in the context of network PC provided to an Enterprise for supporting Teleworking capabilities.

The major features of the service are:

- elasticity in allocation of processing and storage capabilities as a consequence of the dynamic needs of running applications;
- migration of the virtual terminal based on the actual location of the corresponding physical mobile device.

Also in this case, the virtual terminal capabilities can be packaged as PaaS and/or SaaS offering.

5.3.5 An Agenda for Telco-Oriented Cloud Platforms

As seen, web companies have an edge from the technological and the market perspective over the Telcos. In order to recover the gap, there is a need of a coordinated and standardize set of actions aiming at the definition of an open, programmable, and federated cloud platform.

Following the offering of the web companies using proprietary platforms (maybe acquired by IT companies) does not solve the problem. Also, IT companies are lacking behind and too many proprietary solutions do even fragment the cloud computing market. Operators should take actions in order to move toward a grid-based approach. This means to embrace the heterogeneity of the platform components and the ability to mix and match resources pertaining to different administrative and technological domain. The task is more complex that aiming at the construction of a single and proprietary solution, but the risk is to be kept at the margins of an ever-increasing market.

In addition, virtualization capabilities are emerging also in the realm of sensors and smart objects as well as in the networking itself. This will configure a situation in which the dynamic combination and programmability of processing, storage, sensing, and communication resources will move intelligence, services, applications, and infrastructures toward the edge of the network and within the customers' domain. The Telcos can offer to these edge environments programmable and on demand connectivity as well as the ability to support mechanism for self-management of complex edge networks as well as complement the locally missing resources with virtualized ones. This is a daunting objective and it should be approached with a step-to-step strategy. The starting point is interesting from a technical and marketing point of view: the integration of Enterprise IT systems within a federated cloud approach. From a marketing perspective, this means to create a sort of hybrid cloud in which the enterprise IT systems not only maintain their importance but can also cooperate with remote systems and access to specialized applications in order to improve and enrich their functions. Such a hybrid cloud platform could become a sort of cooperative environment in which different enterprises can implement or integrate companies' processes, functions, applications, services, and market places for conducting business.

The cooperative cloud is then instrumental to create an ecosystem in which different enterprises contribute in terms of resources and final customers can access a large set of specialized services.

In order to make such a platform a reality, Telcos should be instrumental to the definition of open architectures within important standard bodies. The more time elapses without this standardization effort, the more the cloud computing business is in the hands of web companies proposing walled gardens.

Some core specifications already exist and they are in the field of Open Grid Forum and Web Services. Telcos should somehow endorse them and to promote a further level of standardization and visibility. The concept of "Virtual Organization" behind the grid computing fits well with a possible business proposition aiming to support the dynamic needs of a virtual organization (e.g., many small companies that dynamically associate in a larger organization for a specific business goal and for a limited period of time). Telcos should pursue and strive for providing a flexible platform capable of accommodating the changing requests of business customers. In addition, support to network virtualization (e.g., OpenFlow) should be guaranteed. Software-defined networks are the next step toward a change in the connectivity and communication proposition of operators. The possibility to create virtualized views on network capabilities and to offer them in an integrated manner with virtual private networks or even better with virtual organizations will provide a viable business framework for the operators. A large ecosystem of application developers, process system integrators, and IT companies could exploit the capabilities offered by such a different cloud platform.

5.4 Bank of User Data

Personal data are gathered and exploited by many WebCos in order to support their Business model (usually related to advertisement). Data are collected in several ways, e.g., by monitoring the activities of users (e.g., cookies), by storing the information they put in their profiles (e.g., Facebook) or by storing and associating the search history to the account of users. There is a continuous trend to get more and more data from users and then to derive information (and possibly sell it) by applying Big Data and Business Analysis related technologies. The Users are constantly spoiled in their rights and property. However, there is an increasing awareness of the importance of the issue, for instance (Benkler 2006) has inspired a group of students of the New York University to develop and run a competitor of Facebook called Diaspora (Fig. 5.13). This awareness is leading to a new approach in dealing and using the personal data (World Economic Forum 2011, 2013).

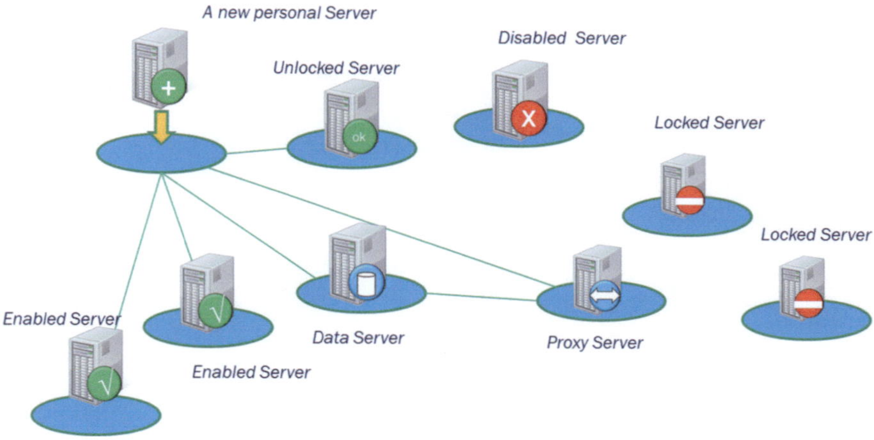

Fig. 5.13 Diaspora, a social network respectful of personal data ownership

For instance, the Diaspora system is characterized by the fact that data are and remain solely propriety of the users. Users can also decide if personal data are to be stored and shared in general servers, or if they have to remain and be shared directly from the user devices.

5.4.1 *The Value of (User) Data*

This is an example of how to deal with data under the push of privacy requirements of aware users. The World Economic Forum (2011) considers the personal data as a big opportunity for the digital economy. Its proposition is to engage the user in such a way to get the usage permission in a more aware fashion.

However this is not enough, the personal data should be dealt with as the "money" of users is treated. The metaphor should be the one of the bank, a bank of the user data. The foundation of this is that data are owned by the user, and they have to be protected and made available to the user whenever it asks for that. If the user wants to invest some of its data, i.e., s/he wants to allow others to use personal data, it can ask the Bank to find out a possible market for selling the data and to state the conditions under which the user will grant the access to its data. In other terms, the bank of user data can act as a sort of Broker that guarantees the investment to its client. Data will be processed, stored, and manipulated according to the rules and requirements decided by the user and not by the providers. Under this condition, a return of investment could also be negotiated between the user and the Broker or the final exploiter of the data. In addition, personal data could also be made available in an anonymized fashion and the Broker could play the role of aggregator of meaningful data sets, guaranteeing on one side the privacy of the users (by anonymization) and the return of investment, and on the buyer side, it

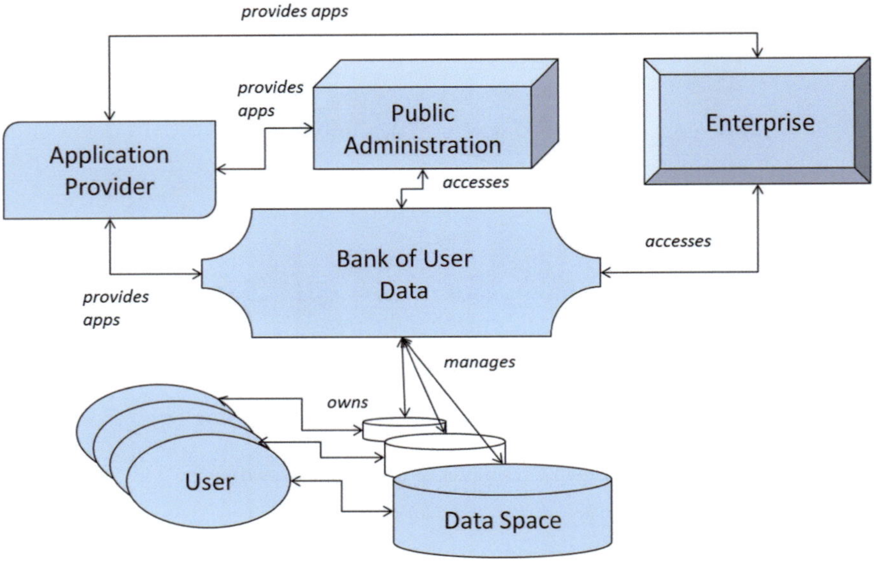

Fig. 5.14 The bank of user data

guarantees the statistical relevance and aggregation of data. Figure 5.14 depicts a possible ecosystem around the concept of Bank of User Data.

Some simple applications could be considered as examples of the valuable features returned to the user:

- The user can have access to the full list of its digital expenditures and in addition it can get information (type, model, warranty) of the products that has been bought. In case of a fault of a product, the user can have an easy way to find out all the documents for a reimbursement or the repair assistance under the warranty. In addition, the user can collect expenditures from several sources and not only from its bank or credit card provider. So its view will be a holistic one.
- The user can provide its anonymized or clear data in return for money or for free services. The contract is agreed considering also the requirements and the policy set forth by the user.

Pushing this approach to the extreme consequences, the user interactions with digital services (even the most common ones) should be transactional, i.e., for any request of the user to a server or a service, a transaction together with the results of the query (in a client–server model) should be returned in order to monitor and collect all the relevant information from the user side. Figure 5.15 depicts this occurrence in the case of a search query. It also shows that the profile of the user should be built by the user itself.

The value of this approach is that the user could collect a holistic view of its actions, behavior, and habits. Extending the example in Fig. 5.15, the user could allow the search provider to use a relevant part of the user profile in order to better

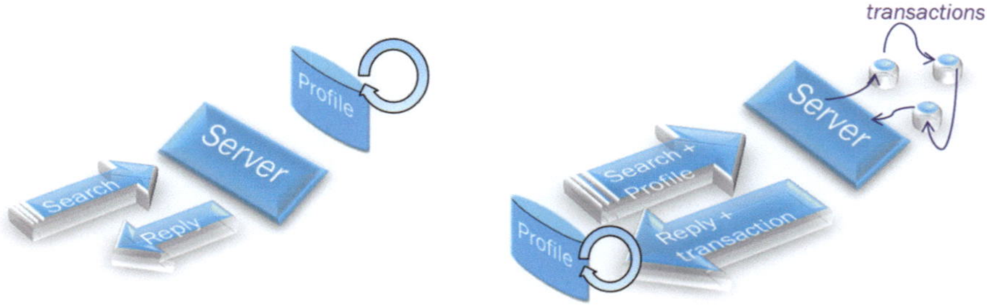

Fig. 5.15 Adding transactions to the web interactions

satisfy the user request. The user profile could contain a list of interests of the user, a part of the aggregate search history (spanning over several search engines), and the like. The service could be more scoped down to the real interest of the user and the collection of responses and transactions could further enrich the user profile to be used for future interactions. Only the user will have access to the whole set of personal data. Entering in this transactional Internet will be more and more important for empowering the user over merchants and service providers. More information on the Bank of User Data can be found in (Minerva and Crespi, Unleashing the disruptive potential of user-controlled identity management 2011; Moiso and Minerva 2012).

5.4.2 Toward an Architecture for Dealing with Aggregated Data

In order to capitalize the value of data, there is a need to create an ecosystem supporting meaningful value chains. The approach advocated is the one termed "User Centric"; i.e., the data generated by users are owned by users and they could be managed by other parties if and only if the user decides to share them according to precise policies negotiated with the other party. These relationships should be supported and made possible by means of an enabling platform. Since data pertain to users, in principle data can be scattered over many different user systems. For this reason, the usage environment has to be intended as a highly distributed platform able to track and integrate distributed data. Figure 5.16 represents a logical data management platform.

Data will be collected in different fashions, some data can be directly provided by the data generators, other data can be collected by observation of activities or behaviors of devices and users, other data can be retrieved by accessing data offered by service providers (e.g., Telcos or web companies), finally other data can be semantically derived by the web and specific data sets made available by environments or communities. One first important level is represented by the set of

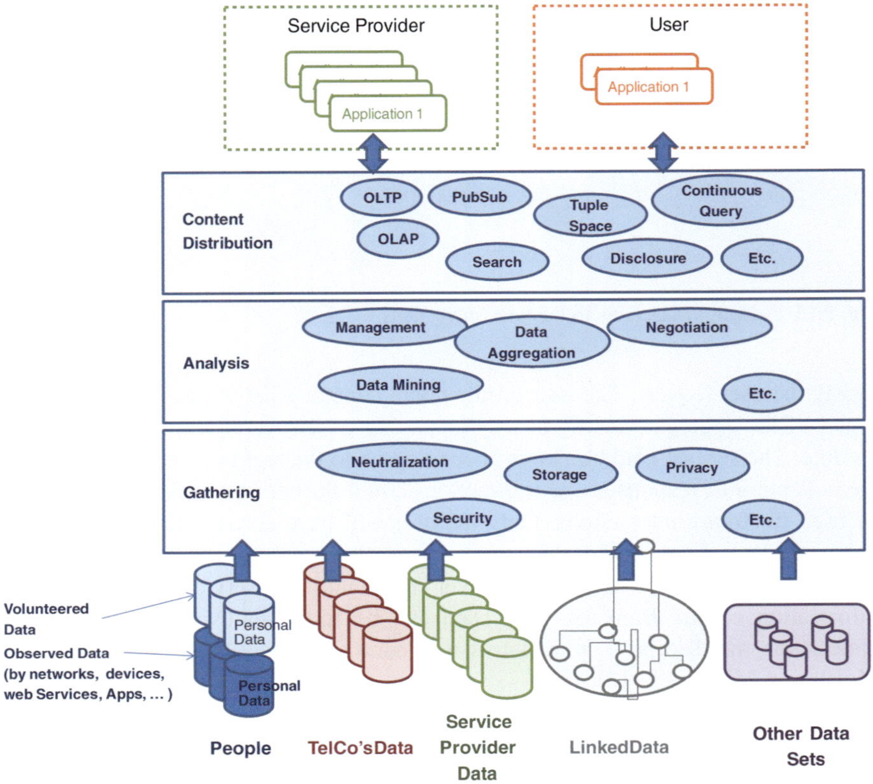

Fig. 5.16 An architectural framework for a data-centric ecosystem

functions that allows the collection, gathering, and organization of raw data. At this level, it is important to provide good mechanisms in order to support capabilities, but even more important are those mechanisms that allow the neutralization, the privacy, and the security of data. Once data have been gathered and organized, the second level is related to the analytics of the available data, here several functions and mechanism can take place. Data Mining is the primary function, but also the possibility to correlate at real time the data is important (e.g., NoSQL techniques). The upper layer is supporting different application-oriented mechanisms for accessing the available information. Access to these data and information will follow different paradigms such as: transaction oriented, i.e., data can be inserted, or read, or modified, or deleted in a single transaction; stream oriented, i.e., data and information are produced and are continuously forwarded to the consumers as in a stream [in this case, mechanisms such a PubSub (Tarkoma 2012) can be offered]. Other mechanisms can be offered such as tuple spaces and blackboards or continuous query. It should be noted that data availability and usage is changing, for instance, there is a trend in considering new way of dealing with data. Streaming computing [as proposed by IBM (Turaga et al. 2010)] is an interesting case of how

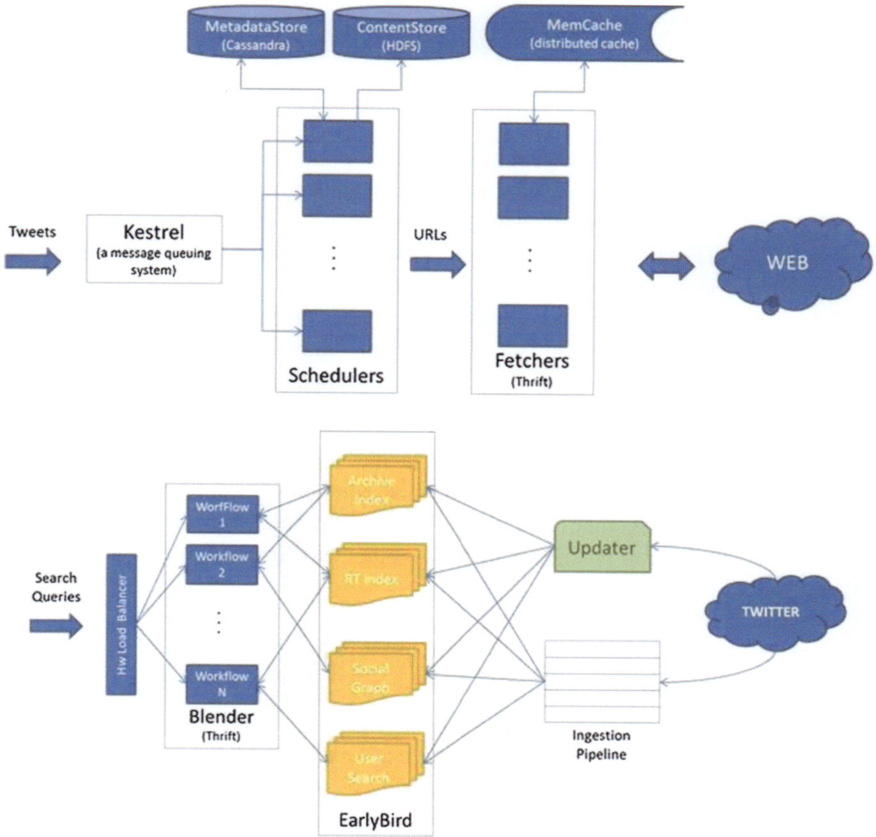

Fig. 5.17 Blender and SpiderDuck, two building-blocks of twitter technology

to process streams of data using a chain (or a graph) of cooperating nodes. The data streaming processing [e.g., S4 from Yahoo (Neumeyer et al. 2010), Deduce from IBM (Kumar et al. 2010) and in general, complex event processing solutions (Buchmann and Koldehofe 2009)] are paving the way for new paradigms for dealing with large sets of data.

The platform depicted in Fig. 5.17 should be capable of supporting several mechanisms to deal with data. One interesting concept that will be more and more important is the need to store the data and to make them available by publishing or forwarding them according to the needs of the applications. Many technologies can be used, but the concept of "store and share" is fundamental in order to enable compelling applications.

In this field, it is worth to mention the Twitter architecture (Krishnamurthy et al., A few chirps about twitter 2008) that could be considered the best implemented example on how real-time data are stored, processed, and made available to the users. Figure 5.17 depicts two important components of the Twitter architecture.

The Twitter architecture is mainly based on open-source technologies that the web company integrates in an innovative way in order to deal with real-time data. To the best knowledge of the author, there is not a clear and complete description of the Twitter architecture, but mainly a description of its components. For example:

- Blender: Queries from the website, API, or internal clients at Twitter are issued to Blender via a hardware load balancer. Blender parses the query and then issues it to backend services, using workflows to handle dependencies between the services. Finally, results from the services are merged and rendered in the appropriate language for the client.
- SpiderDuck: it is a service at Twitter that fetches all URLs shared in Tweets in real time, parses the downloaded content to extract metadata of interest and makes that metadata available for other Twitter services to consume within seconds.

Twitter is particularly important because its capability to deal with real-time stream of data can be used to develop very compelling applications in several domains. In the future, the Twitter "engine" and its architecture could be more important as application enabler than the Google search platform and it is using a different paradigm than C–S.

5.5 Smart Environments

A smart environment is "*a physical world that is richly and invisibly interwoven with sensors, actuators, displays, and computational elements, embedded seamlessly in the everyday objects of our lives, and connected through a continuous network*" (Weiser et al. 1999). Smart environments will be fostered by societal efforts (e.g., Smart Cities), commercial offerings (transportation, home automation, …), and end users endeavors (e.g., www.bwired.nl, or even FON[4]).

Smart Environments integrate the pervasiveness of processing/storage/sensing resources with intelligent behaviors and the ability to understand the "context." Smart environments will be characterized by:

- the intelligence of the ambient, i.e., the capability of the system to analyze the execution context; to adapt the ambient resources and the user devices in order to accommodate the user expectations; to learn from user and resources behavior;
- the awareness, i.e., the capability to identify, locate, and orchestrate resources according to the identified users of the ambient (the humans) and their understood needs and intentions.
- The ability to melt data stemming from different sources in order to virtualize or augment the user perception of the surrounding environment.

---

[4]http://corp.fon.com/.

The concept of context is central to the smart environments. The user context is made out of different components/resources of various networks that are integrated in order to create a single environment that fits the dynamic user requirements. There are a few typical characteristics of the user context such as:

- Physical context: i.e., lighting, noise, traffic condition, temperature, and the like;
- Time Context: such as time of a day, week, month, season of a year;
- Computing context: it is defined in terms of processing elements, storage capabilities, network connectivity, communication cost, communication bandwidth, nearby communication resources;
- User context: it is defined in terms of user profile (the collection of preferences and data related to the user), location, social situation;
- Emotional context: it should encompass feelings, psychology…
- Social Context: in terms of People, Conventions, Etiquette, and so on.

A few of these characteristics are referring to time and space or to physical objects (e.g., the computing context), while the most difficult to grasp and understand context aspects are related to users and their personality and social links and behaviors. Context awareness (Baldauf et al. 2007) is one of the examples of the coordinated work that different sciences have to carry out in order to achieve meaningful results. Cognitive aspects will have an increasing relevance in here, in order to understand the human behavior and the intentions, for creating new compelling interactions modes, for "injecting" intelligence within computing environments. Actually, three main trends in cognitive studies can be identified in ICT:

- Cognitive architectures (mainly related to Artificial Intelligence (AI) studies) aiming at providing reasoning and adaptation to changing situations;
- Cognitive networks (e.g., Clark et al. 2003) aiming at providing a cognitive view on networked resources. They work toward a knowledge plan for the Internet;
- Autonomic and self-organization architectures, aiming at the ability for each node to self-organize and work in complex environments by means of a control loop.

A cognitive approach has been extensively applied to specific problem domains (reasoning about a situation) but just recently it has been proposed for specific problem domains AND the supporting infrastructure [especially in FP7 projects and Internet of Things (IOT-A 2011), iCore[5]]. This means that the cognition will be applied to humans as well as to environments. In fact, in the long term, Cognition and Intelligence for Digital Ambients will emerge as a viable technology.

[5]Documentation available at http://www.iot-icore.eu/public-deliverables. Last accessed April 9th 2013.

Many European projects are putting considerable effort in this area aiming at easing the Self-CHOP management of systems of increasing complexity:

- Self-CONFIGURING, i.e., a system can dynamically adapt to changing environments
- Self-HEALING, i.e., a system can discover, diagnose and react to disruptions
- Self-OPTIMIZATION, i.e., a system can monitor and tune resources automatically
- Self-PROTECTION, i.e., a system can anticipate, detect, identify and protect against threats from anywhere.

Availability of these solutions at industrial level will radically transform the way ICT systems are managed and operated, with great impact on OPEX and redefinition of internal processes as well as a great advantage for users that will be able to relay on smart environments able to adapt to real needs.

Smart Environments will be built around the combination of four major features supported by environment's nodes: processing, storage, communication, and sensing/actuation capabilities. In principle, the combination of these features plus their spatial extension (e.g., small-scale communications, small nodes like RasperryPi (Upton and Halfacree 2012) versus the use of large-scale communications offered by operators plus the access to large datacenters) will determine the effective range of usage of the smart environment. They could range from very small and specific environments in the home, in a lab, up to large environments covering an entire city (smart city) or an entire nation. However, even small smart environments could be interconnected in such a way to create a sort of "network of networks" whose capabilities and whose extension can go beyond the local and small extension. Using a combination of store and forward capabilities, data generated in a local and smart environment could be transmitted in a hop-by-hop manner exploiting opportunistic connectivity. Data could be downloaded and stored in a car and then their destination tagged. The car traveling in the countryside could act as a sort of DHL carrier. Simple calculations have been done for determining the best way to forward large bulks of data, and the use of carriers like Fedex is better than the use of networks. In a way, cars, trains, and ships in smart environments could be cheap and available means to transfer in a "social" way large amount of data. When a well-connected spot is reached, these data (or a part of them) could be forwarded by the car to the new actor by means of the available network. In areas that are densely populated, the small node connectivity could in principle be a sort of pervasive fabric of connectivity capabilities offered by a great variety of devices. Data can "float" in this environment or can travel through it in order to reach destinations far away from the source. The entanglement of data (Minerva and Crespi, Unleashing the disruptive potential of user-controlled identity management 2011) is based on similar concepts. Data could be stored and made permanent even if the nodes of the P2P network offing this service are not. Each node could store and host a copy of the data and replicate it over other nodes. If not all the nodes are turned off simultaneously or in a short period of time, there is the possibility to make the shared data persistent.

Another interesting capability is that all the transport links could be based on small-range connectivity between close and adjacent nodes. This means that a "network of networks" could represent an alternative service to large-scale public network for certain applications (not real time and with simple QoS expectation).

In order to achieve this interconnection, a sort of social "factor" should be present and exploited. As previously seen, one of the problems in similar environments (e.g., P2P networks) is the opportunistic behavior, i.e., a set of nodes take advantage of the features and capabilities offered by the network without returning any value to the community. There are many studies tackling the problem to enforce or at least to promote an altruistic behavior. In Ohtsuki et al. (2006) such a behavior occurs when the number of links between a node and its neighbors exceed the ratio between the benefits/cost associated to the sharing of resources. In a jam-packed environment, the occurrence of altruistic behaviors should be facilitated and promoted by the large number of connected nodes. Nodes of such a network of networks could be characterized as social nodes, in the sense that they are aware of their importance for ensuring the consistency of the network of networks.

These smart environments could be characterized by these features:

- smart environments could be seen as new communication entities (a kind of new terminal/network);
- smart environments can be composed by: (a) simple objects that are controlled intelligently by the user device; or (b) smart objects that can cooperate and offer intelligent functions.

In any case, the aggregation and integration of these environments will present hard to dominate issues. They are similar to those exposed by complex systems. In this sense, strategies for optimizing the usage of resources cannot follow traditional approaches and have to integrate dynamic networking features (as those supported by P2P networks) and autonomic and cognitive ones in order to enforce an intelligent behavior to the whole system. Terminals will play a major role in providing this intelligence, however, the operators could offer optimization functionalities at the edge of their networks in order to help in coping with the complexity. Actually, the public network edge nodes could behave as anchor points in order to introduce some linearity and stability to very dynamic systems.

From an application point of view, end users will be fundamental actors of smart environments: their increased awareness for social and environmental issues (e.g., reduction of CO carbon footprint, energy savings, and the like) and the willingness to take advantage of technological evolution for easing life will create opportunities for the development of these smart environments and especially for applications and services. From the implementation and deployment points of view, social communities will be fundamental for giving a viral push toward the use of smart environments and solutions; developers communities as well could help in order to reduce barriers to deploy and use smart applications that integrate, make use and support smart objects. Smart objects will be more and more available thanks to low-cost sensors and actuators and by the increase capability of end users to build smart objects themselves. Programmability issues will also lessen because of new specific solutions or programming languages for those kinds of environments [e.g., the Scratch language (Maloney et al. 2010) for kids is able to control sensors or AmbientTalk for dealing with not well-connected environments (Dedecker et al. 2006)]. As demonstrated by a number of success stories, crowdsourcing for developing hardware and software solutions tailored to customers' needs could find its way as an established way of promoting new products or elicit the user requirements.

From the communication stand point, it should be noted that mobile terminals are already offering (and they will support even further this option) the possibility to integrate sensors (from NFC up to more complex ones). This feature together with the continuous development of low cost or cheap pervasive systems, gives rise to the possibility to use various communications means (direct communication, Wi-Fi, Zigbee, mobile wireless) for interconnection of different environments. In addition, there are already some communication engines (e.g., Pachube/Cosm, but also Twitter) that support new paradigms of communication well beyond the client–server. The elements for creating compelling services and to personalize these for the specific customer on top of smart environments are technologically emerging. The major issues have a nontechnical nature: how to accommodate for open ecosystems supporting this complexity, what are the advantages for companies to adopt this approach and the like. Once more, the service offering will be determined not by the fulfillment of the customer requirements in an open market, but by the resolution of tussles within an ecosystem dominated by proponents of closed environments (e.g., the web companies, the terminal manufactures, the consumer electronics industry, the operators).

Smart environments will benefit a great deal from data manipulation techniques (e.g., big data) or from graphical and semantic technologies (to be used to present to the user the right information content).

5.6 Internet with Things

Internet of Things is a "catch-all" name referring to several technical and business possibilities. Usually, Internet of Things is seen as a set of vertical application domains that share a limited number of common functionalities. In this document, the undertaken approach is to identify a number of common functionalities that can be used to define a common platform. In order to set the problem domain, this definition is adopted: "The Internet of Things, IoT, is a self-configuring and adaptive complex system made out of networks of sensors and smart objects whose purpose is to interconnect "all" things, including every day and industrial objects in such a way to make them intelligent, programmable and more capable of interacting with humans." Plenty of opportunities and technical challenges can be envisaged in this sector.

5.6.1 *The Value of the Internet of Things*

Internet of Things is rewarded by many as a fruitful application domain in order to generate revenues. Many operators are trying to capitalize this possibility by means of initiatives under the broad umbrella of M2M, applications. They are essentially characterized by the fact that a set of sensors can use an aggregator device equipped with a communication board and a SIM. Access to a mobile network is used in order to dispatch relevant data to an application center. Often there is not a need for full mobility (e.g., a device controlling the working of a lift does not really need to be mobile), however, the low cost of mobility boards and their convenience for sending data are considered as facilitators for building these M2M applications. There are a number of issues that are hampering a wide development of M2M applications even from the perspective of the operators:

- The revenue generated by sending a few bytes of data is not considered an interesting business proposition.
- SIMs should be activated/deactivated at will by the service provider without the intervention of the Telco, this means to give up the control of an asset that is considered extremely valuable by operators.
- M2M applications have showed that traffic patterns can have important consequences on the signaling networks, e.g., when container ships approach an harbor, there could be spikes of traffic due to the simultaneous attempt of thousands of SIMs (one per container or more) to connect to a mobile network (Shafiq et al. 2012). This sudden increase in signaling traffic can cause problems to networks without providing considerable revenues.
- Management functions of these type of SIMS (e.g., billing, recharge of credit and the like) are sometimes more costly than the revenue generated.

One major point is that the data traffic for these services is not usually sufficient to justify a huge involvement of operators in this domain. For instance, Fig. 5.18

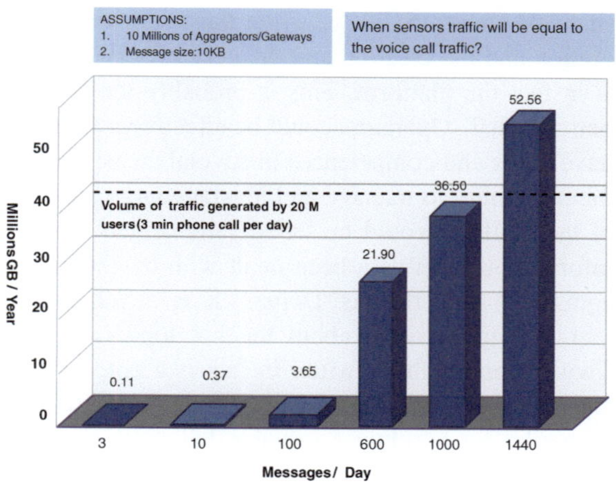

Fig. 5.18 Examples of traffic generated by sensor aggregators

represents the amount of traffic generated by aggregators (i.e., those devices/systems that receive data from many sensors, and manipulate them and deliver the aggregated data to a service center). The assumption is to keep the size of a single message reasonably small (Fig. 5.18 shows data for messages 10 kB long) and to consider a large (10 Millions) base of aggregators. If an aggregator forwards few messages (from 3 up to 100 messages per day) then the generated traffic is less that 4 % of the traditional traffic generated by phone calls (over a customer base of 20 Millions of phone users). Obviously, if the rate of message forwarding increases then also the traffic generated will grow. With more than 1000 messages per day (1440 messages per day means one message per minute) the two types of traffic are comparable.

In the future, it is envisaged that both the message rate and the number of aggregators will grow, but in order to have a substantial traffic increase, sensors and aggregators have to deal and manipulate multimedia information. This means that the size of a single message has to increase (e.g., a photograph) or the single message has to change into a stream of data (i.e., camera recording). So also for sensors, the traffic pattern will drastically change when devices will deal with multimedia data. On the other side, connectivity is becoming a commodity and the operators cannot base their plans in the area of Internet of Things on future traffic increases. There must be something more substantial to this.

The Internet of Things is based on the possibility to measure phenomena and to deal with large data sets. The major value is in the ability to deal with data. The data management has several aspects in this case: the ability to collect data, store them and forward them to the right recipients, the capability to interpret streams of data and to derive relevant information, to aggregate several sources of information and integrate them in meaningful ways. An exemplification of this new way of dealing with massive data is represented by the Fourth Paradigm (Hey et al. 2011).

How to gain this value is a major topic for leveraging Internet of Things. Figure 5.19 represents a hypothesis on how the value of the market of IoT will be segmented.

A traditional business model focusing on communication can aim at getting a share of about 15–20 %, while the major value resides in the platform, its integration and the creation and delivery of services. The major single source of value is in fact the platform. This is probably the starting point for any successful initiative in IoT. Operators could be effective actors of an IoT ecosystem because they have skills and competences in several areas: the communication, the platform and the possibility to manage and leverage large systems (Service Provider). The value of the platform could be further augmented by the capacity to collect and derive information form data been dealt with by the system itself. An example of leveraging real-time data is Twitter. It is a valuable platform capable of providing real-time information about the hot topics discussed in the Internet and in the Twitter community. Twitter in fact is a potential infrastructure capable of seizing and leveraging the Internet of Things value. In the next section, some high-level principles for building a valuable IoT platform are discussed.

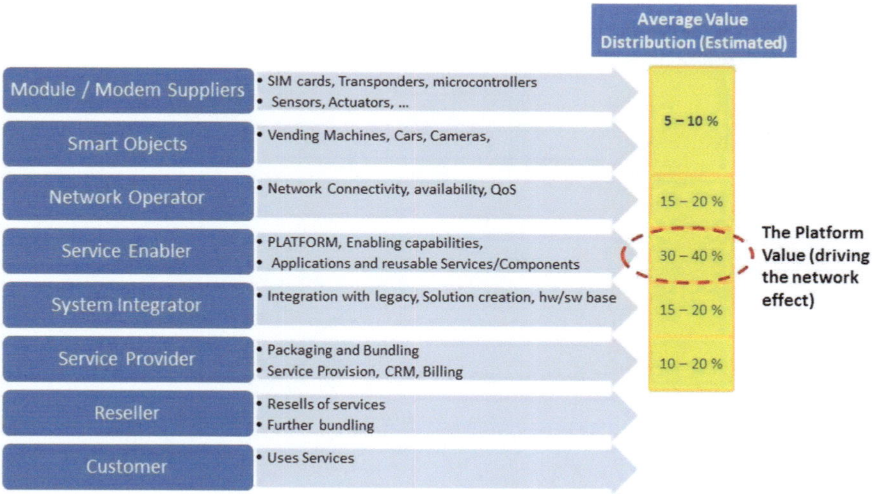

Fig. 5.19 Segmenting the IoT market value. *Source* Nokia Siemens Networks

5.6.2 *Toward a Platform for the Internet with Things*

A viable proposition for the operators in the realm of Internet of Things are large-scale systems, i.e., those systems that comprises thousands and even millions of sensors and actuators devoted to provide useful services to customers and citizens. "Smart cities" will certainly follow in this area. These systems will be characterized by the fact that sensors and actuators will be deployed in a very irregular pattern, i.e., certain areas and environments will be covered with plenty of sensors, while others will have a scarce deployment of them. In addition, sensors will pertain to several administrative domains. This means that complexity will lay in the sheer number of sensors, in the uneven deployment, and in the variety of administration domains. These features can be coped with by means of virtualization and the willingness of users and providers to share resources (Bohli et al. 2009). In such an environment, the willingness to make available a virtual representation of the sensor will have a paramount importance. Service and infrastructure providers should support open platforms in order to allow the brokering and the dynamic integration of heterogeneous resources. Telcos can be trusted providers that integrate the different systems into a meaningful larger environment.

Virtualization is a means to allow high degrees of adaptation and integration of several resources. Each administrative domain in fact could make use of sensors and actuators based on different technologies, communication paradigms, protocols, and software platforms. In addition, sensors and actuators could be used in order to "substitute" or provide similar functions of specialized sensors (not deployed or available in a specific area). For instance, using cameras with visual detection can

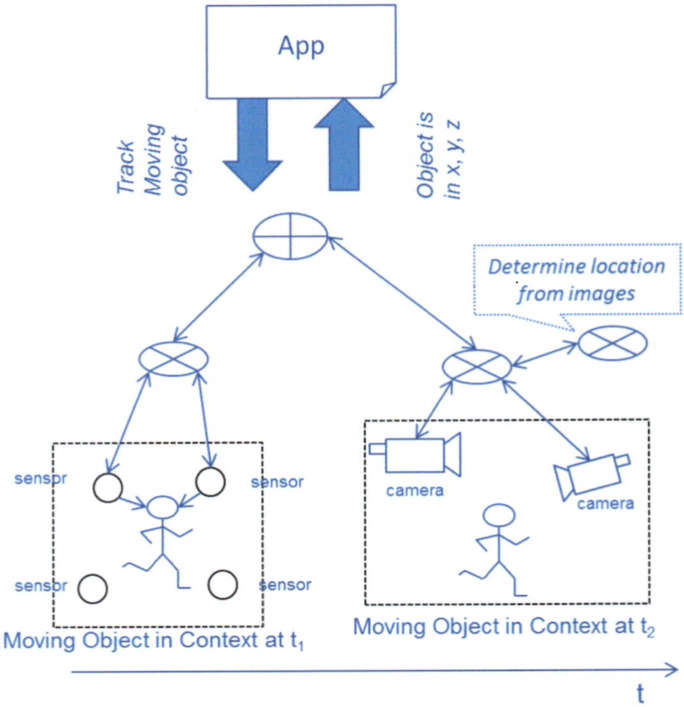

Fig. 5.20 Using real-world objects to "deputize for" missing sensors

be a substitute for location or tracking capabilities as depicted in Fig. 5.20; noise sensors could substitute counting sensors (the level of rumor generated by a crowd can be used to approximately determine the number of users in a location), and the like.

These examples point to the need of integrating sensing capabilities and cognitive behaviors in order to support and open up to a whole wealth of applications. Virtualized objects and their functionalities should be transparently mapped and adapted to provide to applications the expected logical functions. In doing this, the IoT platform should be capable of determining whether available resources and their capabilities can be allocated and whether the available functions can be integrated in order to approximate the logical requested functions. These mechanisms are all based on the comprehension of the execution context and on the adaptation, and integration of functions in order to approximate the needed capabilities. Cognition, reasoning and programming of virtual objects are fundamental ingredients for a viable IoT architecture (iCore 2013). Figure 5.21 represents how to adapt the available basic functions by means of reasoning and situation awareness to the requests of the application.

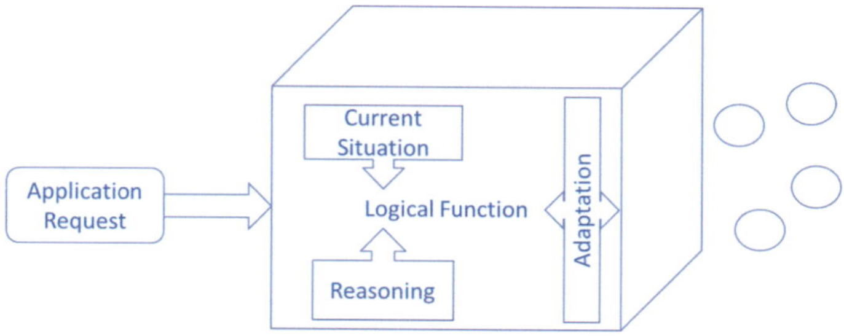

Fig. 5.21 Flexible adaptation of functions by means of cognition

Actually these mechanisms point to a very crucial aspect of cognitive IoT platforms: the need to maximize the satisfaction of requests of many competing applications, and the need to globally minimize the usage of available systems resources. This pushes for two different perspectives (and goals) of the platform: an application view, i.e., the set of functionalities and mechanisms needed to facilitate the creation and the execution of applications; and a system view, i.e., the set of mechanisms and functions that allow the platform to optimize the allocation and the consumption of resources according to availability and requests. It is interesting to note that both these functional views can be supported by cognitive mechanisms, a set for reasoning (and learning) about the functions needed by applications and another set for reasoning (and learning) about the working of the system itself.

There is an interesting capability to leverage in order to reach a high penetration of sensing features: the crowdsensing (Ganti et al. 2011), i.e., the possibility to use sensors in mobile devices in order to collect and provide data to applications. In spite of many technical issues, crowdsensing is an attempt to overcome one of the major issues in Internet of Things: the wide deployment of sensors. For instance, in order to make a city a smart city, plenty of (wireless) sensor networks should be deployed. Even if the cost of sensors and machinery is rapidly decreasing, the cost of deployment of the infrastructure and its maintenance are still high. Extensive deployments require a lot of investments without the assurance of any economic return. Crowdsensing may be seen as a way to involve users in sharing the data they can collect about a specific environment or a city. At the same time, the individual user has to be rewarded for participating to the data gathering. The collection should be respectful of preferences and policies stated by the users, but in any case, users will be a primal source of data about the environment.

As seen, data gathering will be very heterogeneous and data themselves will be collected in different formats. There is a stringent need to "normalize" these interfaces, layers and data formats in order to create a unique virtual platform on top of which to build applications and services. Figure 5.22 (from iCore 2013) depicts how adaptation and virtualization can help to build such a homogeneous view on data.

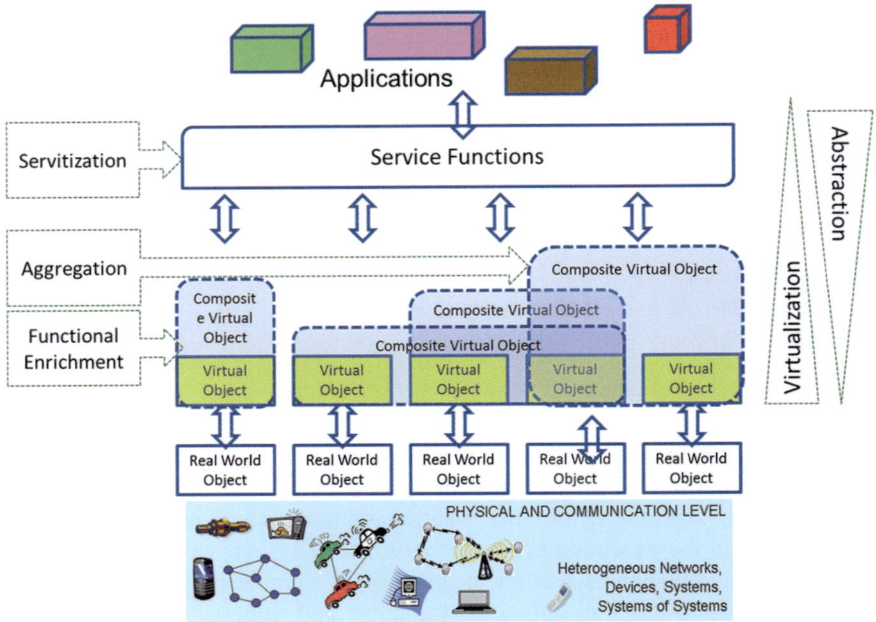

Fig. 5.22 A generic view of an IoT platform (iCore architecture)

It should be noted that aggregation of functionalities leads to abstraction (that generally speaking is a good property for a software system), however, this property has to be carefully used because abstraction means also to have a coarse grain control on resources and sometimes applications need to exert a very granular and specific control over resources' functionalities. The issue of abstraction versus granularity has been discussed in (Minerva 2008). One viable solution is to offer different levels of interfaces and APIs that support different degrees of abstraction. This approach is depicted in Fig. 5.23.

As a simple rule of thumb, it is important to expose for each resource type all the interfaces it is offering as a stand-alone interfaces (not mixing up different interfaces and/or protocols). These basic interfaces can be integrated and mixed if the services require a simplified and aggregated access to the underlying functionalities. At the upmost level, APIs and interfaces can abstract functionalities in favor of simplicity. Simple applications will use simple interfaces. Those requiring a granular control over the specific features of the resource will have the possibility to access them at the cost of dealing with more complexity.

Virtualization, complex systems technologies (e.g., autonomics) and cognition are capabilities that probably will change the meaning of Internet of Things, in fact they allow for a huge transformation: any single object (even physical or immaterial) can be represented in a cloud and being used, mashed up, and transformed.

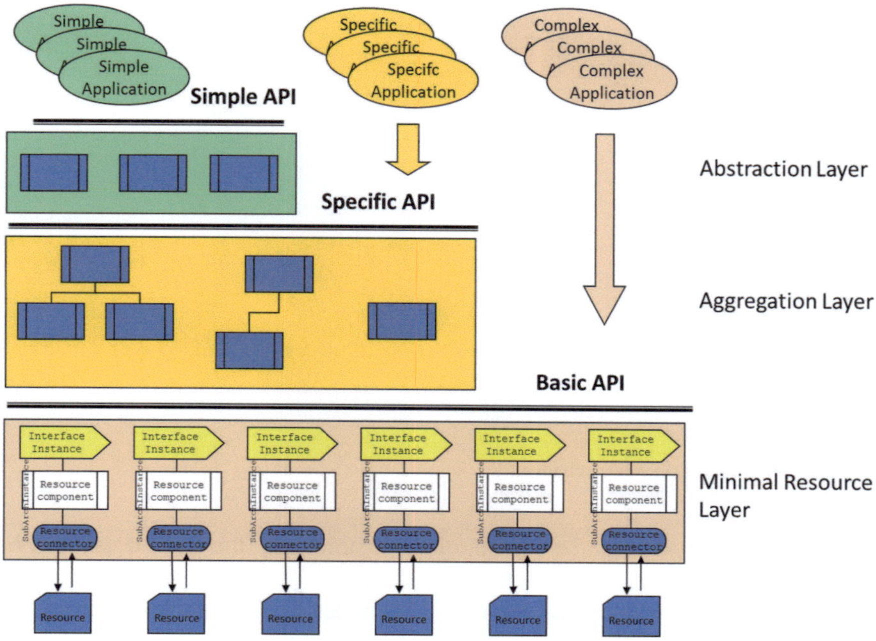

Fig. 5.23 Composition of interfaces and programmability

In this case, it is possible to move from Internet of Things to an Internet **with** Things. To clarify this concept, Internet of Things can be seen as an environment built by the combination of low-cost standard sensors, short-range communication, capillary and macro networks, vertical services, data aggregation in cloud, and the possibility through specialized interfaces of third-party development. The Internet with Things can be seen as a large horizontal environment comprising "Virtual Objects" that mirror "Things" in cloud, which extend objects with semantics, allow extended data integration and aggregation, support federation and portability of different data formats, and uses the cloud as a development platform. These differences enable a lot of new applications and domains. Some of those are sketched in the next subsections.

Another interesting aspect of the Internet with things is represented by the capability to deal with streams of data in an efficient way. This can yield to the spread of different paradigms beyond client–server. Actually, one communication paradigm, the Publish/Subscribe (PubSub) (Tarkoma 2012), is interesting and is applicable to IoT as well as to social relationships The working of PubSub is strongly based on the message passing paradigm. Recipients subscribe to topics/data they are interested in and senders send messages to the queue associated with the topic/datastream. A broker then forwards messages to the subscribers that can use the receive data. Figure 5.24 represents this communication paradigm.

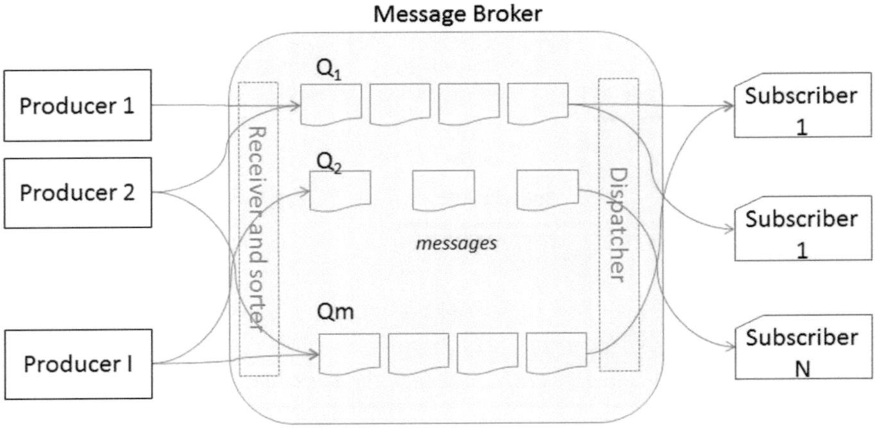

Fig. 5.24 A typical PubSub mechanism

Similar mechanisms can be used in order to store data (the queued data can be sent to a storage function for instance) in order to have an historical trace of the stream of data that applications can use for some analytics or for log purposes. In addition, the messaging system can try to do some real-time analysis on data passed in order to derive some information. For instance in the case of monitoring sensors, if the messaging system receives many alarms, it can try to relate them in order not to overflow the queues or to correlate them so that different causes can be skimmed out.

Actually, this communication paradigm is at the base of Twitter. It is interesting to note that this platform is aiming at sharing information and news within communities, but some developments (e.g., Kranz et al. 2010) are using this platform for providing IoT related applications. Two considerations are important:

- The real-time web (and in particular the PubSub model) could support new classes of services and enabling new platforms and providers;
- The PubSub mechanisms applied to Internet with Things enable the Brokering role for information exchanged by smart objects and the extraction of relevant information.

Figure 5.25 represents a possible extension of the PubSub functionalities that recalls in a way Twitter.

This platform can be considered a sort of Twitter of Things, because it is focusing on the real-time exchange of information and data between sensors and smart objects and a set of applications. In addition, it is a complex event processing engine that can extract knowledge by analyzing the type of data being dealt with. As said, in the Internet of/with Things, a fundamental role is the one of platform provider, this Twitter of Things platform can be quite important for playing this role and the broker one.

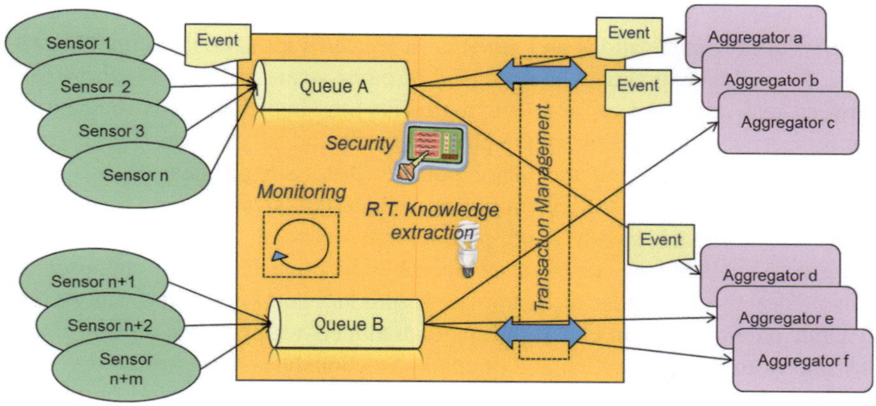

Fig. 5.25 A transactional complex event processing derived from twitter

5.6.3 Virtualization-Enabled Scenarios

Virtualization is a powerful concept that can be used to develop a number of disruptive or at least innovative services. An interesting point is that virtualization can enable servitization, i.e., the capability to transform objects and products into services. On the other hand, virtualization of objects is strongly tightened to Identity Management, i.e., how objects can be related to a user of them and under what conditions and privacy features. Virtual Objects can be created, extended, and mashed up by several users that will share the final outcome, each one owning a part of the composed virtual object and the related environment. Creating a synergy between the virtual continuum and the identity management systems offers the possibility to highly personalize an environment to the needs of a specific customer or its community. In the following, two examples of virtualization capabilities are provided.

5.6.4 Virtual Continuum

Virtualization can be exerted to such a level to allow the representation of almost any physical object in the "cloud." Each clone object could:

- enrich the characteristics of physical devices, in terms of processing, storage, reliability, connectivity, battery usage, etc.;
- break the fences of the walled gardens of device vendors, by creating an alternative/independent service ecosystem;
- enable the virtualization of "physical device ownership," by enabling people to use shared devices/objects, as their private objects (e.g., personalization).

This level of Virtualization is quite interesting from the point of view of the creation of new applications and services. It creates a virtual continuum (Minerva et al., Toward an expressive, adaptive and resource aware Network Platform 2011) between real-world objects and their representations in a digital environment that can be seen as a bridge between the physical world and cyberspace. Objects can be functionally extended, but the major point is that each physical object, e.g., a product, can become a service. This approach, called servitization (Baines et al. 2009), tries to transform any good, product into a service. A product is presented and made available as a set of composable functionalities that the user can activate/deactivate in order to customize the product and to have it to fully satisfy dynamic requirements. In a way, the single product is wrapped in a functional shell that allows to deliver the free and premium functionalities on demand.

The virtual continuum is the constant entanglement between real objects and their representation in the network. Events, actions, data on physical objects will be represented in the virtual world and vice versa. The virtual continuum makes possible the close relation between atoms and bits.

Introducing virtualization in the Internet of Things context could have many merits. For instance, it could help in overcoming the heterogeneity of many proprietary architectures and systems enabling the possibility to run several proprietary IoT applications into a single platform. Virtualization of sensors and actuators is also important because it allows to represent real-world objects into the network. This enables the possibility of representing and programming a real-world object in the iCore platform and to control/govern/integrate the virtual instance in a programmable environment. Virtualization is the first step toward the virtual continuum, i.e., the possibility to create a link between any physical object and its representation in the cloud. The virtualized object can extend the functionalities, the features and the capabilities offered by the real one. For instance, a real-world object, a car, can be virtualized in the cloud and its functions can be controlled and managed in the "Internet." The car functions can be extended by applications and services running in the cloud. The virtualized car can be enriched with control functionalities (e.g., for a trip in the mountains, the driver can enable the four wheel drive capability for a couple of days by interacting with the car and paying for that functionality for the required time. Additional power can be bought by means of the virtualized representation of the car for a specific travel, and so on). The car is not anymore a product sold to a client, but a service that can enable and disable premium functionalities depending on the needs of users. This is an example of how virtualization can support servitization. Different objects could represent a single real object allowing for sharing of its functionalities in different virtual environments. The same physical sensor can be virtualized and framed into different contexts of usage. Valuable resources can be virtualized and shared in such a way to increase their usage and to help in limiting the wide deployment of similar sensors by different providers. On the other side, valuable functions can be derived by virtualizing the real sensor capabilities: e.g., a camera can be used as a location device, a noise sensor can be used to determining the number of people present in a

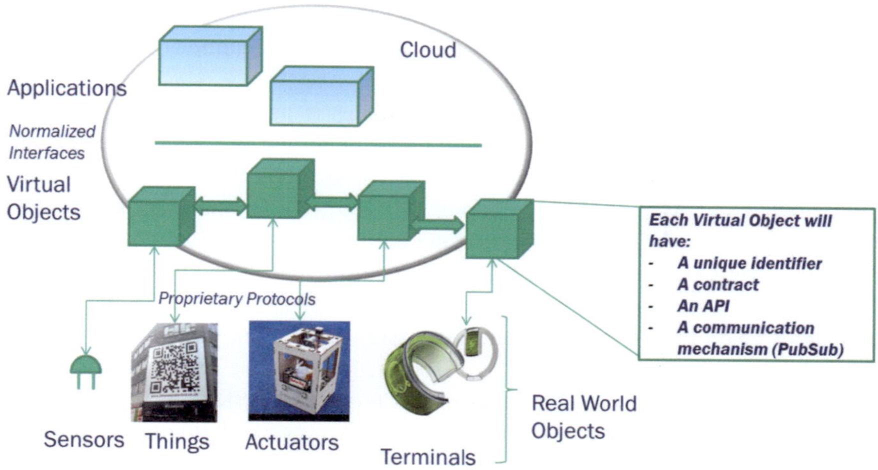

Fig. 5.26 A representation of the virtual continuum

street, and so forth. Virtualization allows to derive usable functions by many heterogeneous devices deployed in the real world.

This virtual continuum (Fig. 5.26) can be seen as a sort of platform that enables the migration of control from the device to the cloud. In this way, the "ownership" moves from the device producers to cloud providers and it allows users to benefit from a larger application ecosystems.

Some features of the Virtual Continuum are:

- Decoupling of physical objects and virtual ones
- Normalized applications can be built on virtual objects (having an impact on physical objects)
- Plenty of applications can be built on top of virtualized and extended objects
- Leverage of cloud infrastructure.

5.6.5 Virtual Terminal

Another example of the possibilities introduced and offered by virtualization capabilities is the virtual terminal paradigm (Minerva et al., Toward an expressive, adaptive and resource aware Network Platform 2011): a terminal can be cloned in the cloud and it can act on behalf of the physical terminal. It can also be "composed" by the integration of several physical terminals. Enabling such a possibility could free the user from the need to run applications within a proprietary and confined walled garden. An example of this trend is given by the Silk solution developed by Amazon: the terminal and the server can dynamically determine (considering the network conditions and the terminal capabilities and status) where is the right place to execute relevant functions. Sometimes the cloud can take over the processing burden, but the processing may also be entirely executed in the terminal.

The concept of virtualization of mobile phones is getting traction worldwide. There are already some initiatives related to the definition of virtual terminals and virtual environments. For instance the clone cloud project within Intel Research Center (Chun and Maniatis 2009; Chun et al. 2011) aims at "clone the entire set of data and applications from the smart-phone onto the cloud and selectively execute some operations on the clones, reintegrating the results back into the smart-phone." Also NTT is working on a virtual terminal (Itoh et al. 2010) with the idea of offloading much time consuming and processing tasks from the terminal into the network.

A virtual terminal is a software feature that provides a perfect mirror, constantly updated, of all the actions, data, and occurrences of a specific terminal or a group of them. A virtual environment is a software feature that allows customers to use an entire (virtualized) computational and communication environment tailored to their specific needs.

The concept behind these projects is very simple: to integrate data and execution environments of the terminal into the cloud, i.e., to provide a functional extension of the terminal capabilities into the cloud (see Fig. 5.27).

The objective of this initiative is to drive the definition of a platform capable to (a) support the virtualization of terminals and other smart objects, (b) to allow the Telcos to play the role of aggregator and broker of virtualized (communication, processing, storage) resources and applications; (c) to allow the creation and management of virtual environments tailored to the specific needs of a customer. Figure 5.28 sketches the problem domain.

Some Examples

Moving the SIMs into the virtual terminal. A user could map one or more physical terminals onto a (subscribed) virtual terminal in order to synchronize them or to extend their capabilities. The user decides how to deal with communications and

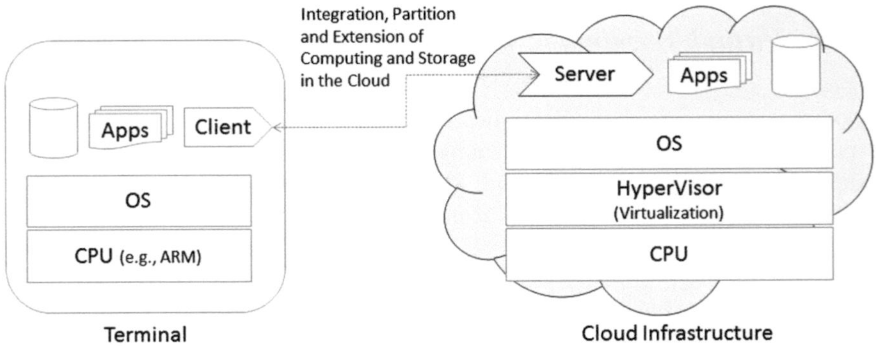

Fig. 5.27 Virtual terminal as a functional enrichment of a terminal

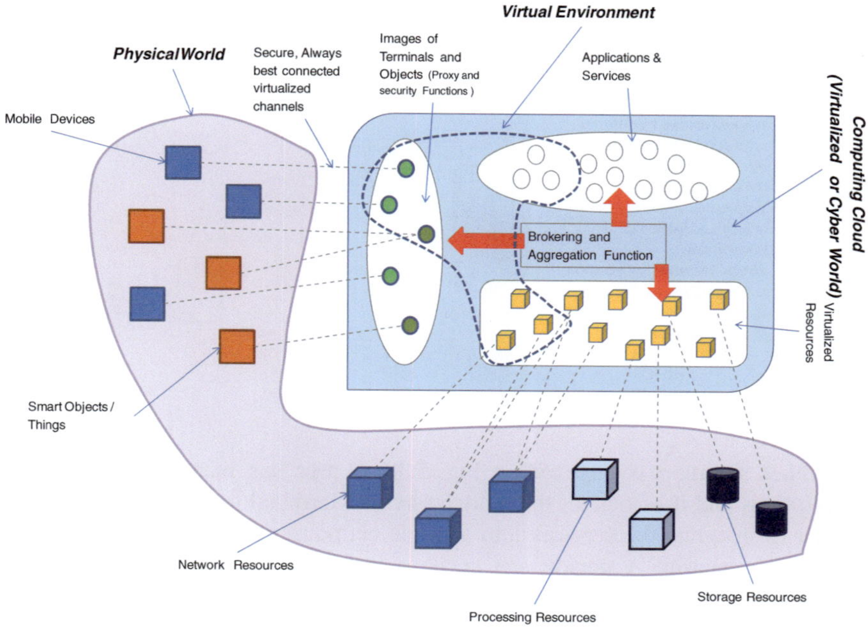

Fig. 5.28 Towards a platform for virtual environments

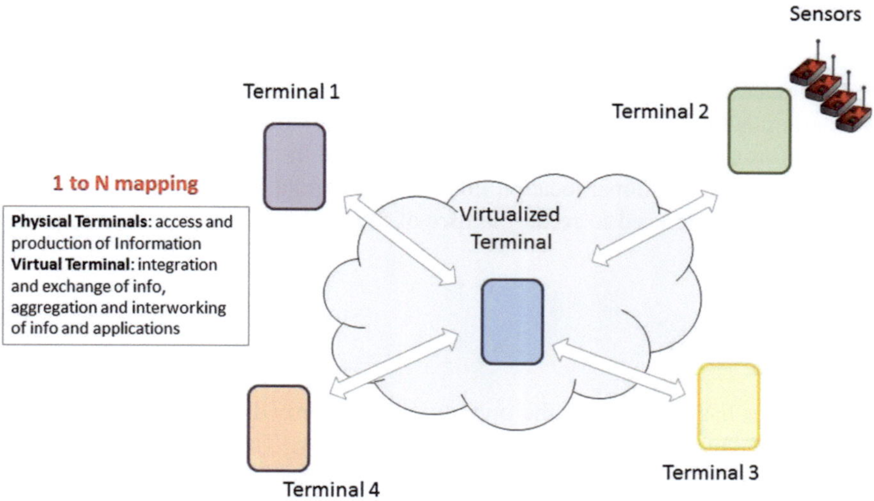

Fig. 5.29 1-to-n mapping

processing services, however, the different terminals can create a sort of mesh network and can partition information and computational tasks among them. See Fig. 5.29.

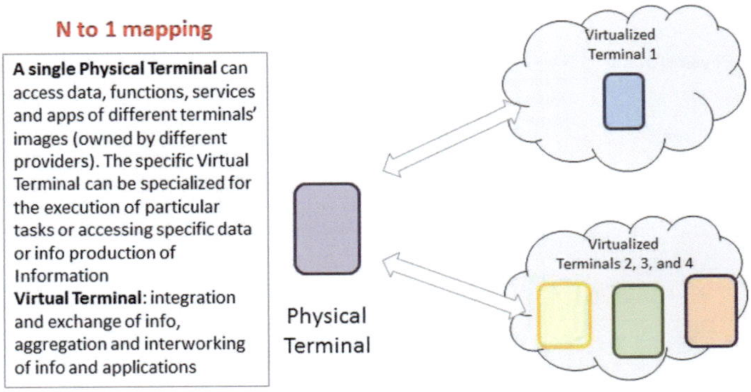

Fig. 5.30 n-to-1 mapping

Another situation is depicted in Fig. 5.30: a user has many virtual terminal subscriptions and it associates the unique physical terminal to all its virtual images in order to integrate the services into a single endpoint.

My trip scenario: A person can create a virtual object called "my trip" that represents all the needed information related to a travel. This object is created in the virtual environment. This object comprises the set of information, data, and alerts needed to make the trip easier or more rewarding for the user. The user first visits the destination virtually by means of a street view and bookmarks some points of interest with alerts. All the booking and tickets are aggregated to the virtual object "my trip" (aggregation of information). At the check-in, the user has just to share with the clerk a pointer to the ticket and gets in return another object (the check-in object) to associate with the "my trip" object. At the destination, the user can start wandering and can ask the navigation support to "my trip" and can get also some useful information about monuments, restaurant, and the like. This instance of "my trip" (plus some additional location information) can be stored in the system. Old information can be used to recall the user of things done or seen in that place.

5.7 Findings

This chapter has presented some opportunities for providing new services. Other services are considered as well by the ICT industry. These services have been considered in order to give feedbacks on traditional services (as the rich communication ones), on current proposals of many Telcos (like the cloud computing) and on possible future opportunities like personal data, smart environments (many operators are starting to work on smart cities) and Internet of Things (all the major Telcos are trying to understand how to move from machine to machine services to Internet of Things). Some of them will be further considered in Sect. 6.5. as examples of new opportunities that "smart operators" can exploit in order to open up and take the leadership in new markets. Other services are newer and deserve more study and experimentation in order to find out whether they are actually viable.

Communication services will not provide considerable improvements with respect to the current situation. The RCS and even the WebRTC based services do not offer anything that is new for the users. They are a collection of already available functions that the Telcos are making more interoperable between their domains. In reality, users have already interoperable communications services, in fact each major WebCo is offering these services. They allow the user to communicate within very large communities (think to Facebook) and interoperability is reached by switching to different applications. Actually, the value in rich communication services is not too much in the communications functions. It is in the community, i.e., how many people can be reached by that application/service. This is an important lesson to keep in mind. Many operators still consider interoperability a major value, but in reality it is not if the community associated to the service is big enough. The revenues generated by these services will not generally be high and possibly they will not even compensate for the investment put forward to create a service delivery solution. Probably, a better approach is to compete in terms of tariffing on existing capabilities.

Many Telcos consider cloud computing and the provision of XaaS functionalities as a major business opportunity, as discussed in this chapter. However, the cloud market is already controlled by other actors (like the WebCos) and the role of operators is restricted to certain types of services and functionalities. A first insight is that there is a need to clearly differentiate the markets: residential is a very competitive one. The Business market is large and is very valuable but it is also demanding in terms of attention of the customers. It is also difficult because customers are more used to determine the best and more competitive offering and they can switch from one provider to the other very easily. In addition, issues posed by this type of customers could be challenging. For example, the local cloud provider could be different country by country. Providing an interoperable federated cloud could be very difficult and the revenues generated by it have to be shared between different partners. From a technical perspective, the adopted solutions will lag behind those deployed and offered by the WebCos, and so customers could continuously ask for improvements.

The newer services offer some possibilities: the Bank of user data can be appealing if a vast majority of user will recognize the need to be protected and not exploited by others that do manage the user personal data. Another viable way of proposing the service is trying to return to the user a part of the revenue generated by selling the user data. An important point that deserves a lot of attention is the identification of parameters and criteria to evaluate the effective value of the data of a single user and the aggregated value of data of several users in the network. However, this service seems to be important because it is based on a real user-centered principle: the data pertain to the users and they have to be managed according to rules and policies determined by the user itself. Even, if there is no business value behind it, this is an important indication that consumer associations and authorities worldwide should carefully consider. As previously seen, personal data and identity are two faces of the same coin and so similar issues related to identity ownership can be put forwards. They will be important issues to be solved at the technical level as well as at economic and regulation levels.

Smart environments seem to be the natural evolution of the integration of Telecommunications and IT. It is important to start very soon in experimenting, providing, and supporting this class of services in order to grasp the user requirements and to provide captivating services. In this area, new interaction paradigms are more important than ever in order to engage the users. In addition, the large distribution of functions and the fragmentation of devices will oblige for distributed solutions adopting new paradigms. For instance, the PubSub one can be a good one in this context in order to forward events to a set of subscriber processes that need to react to the occurrence of an event. P2P solutions are a natural choice for this type of services, but they do not offer standardized solutions and in addition they need a lot of programming. Once more there is a compelling push to operators to enter into different levels of mastering the software.

Internet with Things is the most appealing service scenario for the Telcos. It is a broad opportunity falling over several markets and industries. It has impacts on many processes and for this reason can have an economic relevance for many customers (improving the efficiency of some business processes is a viaticum to success). This application domain allows also to introduce disruption in existing providers and it also can be approached with new interaction and control paradigms. In other terms, it is a market in an initial state and the provider that will identify and provide the best technologies has a good chance to be an important player in a rich market of the future. Telcos can use it also instrumentally for experimenting new paradigms that displace the WebCos (e.g., P2P propositions) and play as technological forerunners for a first time since a long period. In addition, the Internet with Things has a large Business market characterization that can be exploited by the Telcos in order to start to position themselves with respect to WebCos. The virtual continuum creates also the possibility to virtualize terminals and could offer the opportunity to reduce the power of terminal vendors with respect to operators. Users can be offered virtualized and interoperable environments that free them from strong dependences on specific terminals and operating systems. Once more a user-centric approach could be a viable way to revert compromised business positions in favor of new ones more open and favorable to the users.

Different paradigms have been considered in the services described in this chapter. Some of them are well known and fall in the basic ones (i.e., C–S, NI, and P2P). However, different models are emerging especially in the realm of smart environments and IoT. In these cases, there is a need to deal with events in order to allow a granular control of functionalities to applications and the different stakeholders of the service. In addition, these services will comprise many different resources interacting in several ways with each other. There is a need to organize resources and their interaction in such a way that autonomic behaviors can be enforced for limiting the human intervention. These services need to be supported by cognitive solutions (as those described in Sect. 4.6) that will make simpler to program and use newer service architectures. One of the major finding is that services have to be provided with appropriate solutions that do not require a lot of adaptation of the architecture in order to represent the service interactions, but exactly the opposite: the service architecture has to be chosen according to its capabilities (its expressive power) to represent and support in the easiest possible way the interaction and control needs of new services. Flexibility of platforms and architectures, and the ability to encompass new control paradigms are major advantages of future solutions.

Internet with Things and smart environments can be considered as inflection points in which new technologies open the path to new markets and approaches. These classes of services can be considered by Telcos as means to enter into processes and markets with a winning proposal. The book (Minerva and Crespi 2016) will further elaborate on how to enter into adjacent markets with innovative approaches. In particular, Scenarios in this book will use some of these findings in order to show what roles that can be played by Smart operators.

Bibliography

Armbrust M et al (2010) A view of cloud computing. Commun ACM (ACM) 53(4):50–58

Baines TS, Lightfoot HW, Benedettini O, Kay JM (2009) The servitization of manufacturing: a review of literature and reflection on future challenges. J Manuf Technol Manag 20(5):547–567

Baldauf M, Dustdar S, Rosenberg F (2007) A survey on context-aware systems. Intl J Ad Hoc Ubiquitous Comput 2(4):263–277

Baset SA, Schulzrinne H (2006) An analysis of the skype peer-to-peer internet telephony protocol. In: INFOCOM 2006. 25th IEEE international conference on computer communications. Barcelona. IEEE, pp 1–11

Benkler Y (2006) The wealth of networks: how social production transforms markets and freedom. Yale University Press, New Haven

Bohli J-M, Sorge C, Westhoff D (2009) Initial observations on economics, pricing, and penetration of the internet of things market. ACM SIGCOMM Comput Commun Rev (ACM) 39(2):50–55

Buchmann A, Koldehofe B (2009) Complex event processing. IT-Inf Technol 51(5):241–242

Chun B-G, Maniatis P (2009) Augmented smartphone applications through clone cloud execution. In: Proceedings of the 12th conference on hot topics in operating systems. USENIX Association, Berkeley, USA, p 8

Chun B-G, Ihm S, Maniatis P, Naik M, Patti A (2011) Clonecloud: elastic execution between mobile device and cloud. In: Proceedings of the sixth conference on Computer systems. ACM

Clark DD, Partridge C, Ramming JC, Wroclawski JT (2003) A knowledge plane for the internet. In: SIGCOMM '03 proceedings of the 2003 conference on applications, technologies, architectures, and protocols for computer communications. ACM, New York, pp 3–10

Dedecker J, Van Cutsem T, Mostinckx S, D'Hondt T, De Meuter W (2006) Ambient-oriented programming in ambienttalk. In: ECOOP 2006–object-oriented programming. Springer, Nantes, pp 230–254

Ganti RK, Ye F, Lei Hui (2011) Mobile crowdsensing: current state and future challenges. IEEE Commun Mag (IEEE) 49(11):32–39

GSM Association (2009) Working group: "rich communication suite (RCS) & rich communications ecosystem (RCE). White paper. GSMA, London

Henry K, Liu Q, Pasquereau S (2009) Rich communication suite. In: 13th international conference on intelligence in next generation networks, 2009. ICIN 2009. ICIN, Bordeaux. pp 1–6
Hey AJG, Tansley S, Tolle KM et al (2011) The fourth paradigm: data-intensive scientific discovery. Proc IEEE 99(8):1334–1337
iCore (2013) Functional architecture. Deliverable D2.3, iCore Consortium, Brussels
IOT-A (2011) Initial architectural reference model for IoT. Project deliverable D1.2, IOT-A Consortium, Brussles
Itoh M, Chen EY, Tetsuya K (2010) Virtual smartphone over IP. NTT Tech Rev 8(7):1–5
Kranz M, Roalter L, Michahelles F (2010) Things that twitter: social networks and the internet of things. In: What can the internet of things do for the citizen (CIoT) workshop at the eighth international conference on pervasive computing (Pervasive 2010). Helsinki Institute for Information Technology HIIT and University of Helsinki, Helsinki, pp 1–10
Krishnamurthy B, Wang J, Xie Y (2001) Early measurements of a cluster-based architecture for P2P systems. In: Proceedings of the 1st ACM SIGCOMM workshop on internet measurement. ACM, Burlingame, pp 105–109
Krishnamurthy B, Gill P, Arlitt M (2008) A few chirps about twitter. In: Proceedings of the first workshop on online social networks. ACM, Seattle, WA, USA, pp 19–24
Kumar V, Andrade H, Gedik B, Wu K-L (2010) DEDUCE: at the intersection of MapReduce and stream processing. In: Proceedings of the 13th international conference on extending database technology. ACM, Lausanne, Switzerland, pp 657–662
Lin M, Arias JA (2011) Rich communication suite: the challenge and opportunity for MNOs. In: 15th international conference on intelligence in next generation networks (ICIN). ICIN, Berlin, pp 187–190
Liu F, Tong J, Mao J, Bohn R, Messina J (2011) Lee: leaf, dawn badger. NIST cloud computing reference architecture. NIST Special Publication, Washington, USA, p 292
Loreto S, Romano SP (2012) Real-time communications in the web: issues, achievements, and ongoing standardization efforts. IEEE Internet Comput (IEEE) 16(5):68–73
Maloney J, Resnick M, Rusk N, Silverman B, Eastmond E (2010) The scratch programming language and environment. ACM Trans Comput Educ (TOCE) (ACM) 10(4),article no. 16
Minerva R (2008) On some myths about network intelligence. In: Proceedings of international conference on intelligence in networks-ICIN2008 (October2008). ICIN, Bordeaux, France, pp 1–6
Minerva R, Crespi N (2011) Unleashing the disruptive potential of user-controlled identity management. In: Telecom World (ITU WT), 2011 technical symposium at ITU. IEEE, Geneva. pp 1–6
Minerva R, Crespi N (2016) Networks and new services: future of telecommunications. Springer, Berlin
Minerva R, Manzalini A, Moiso C (2011) Towards an expressive, adaptive and resource aware network platform. In: Prasad A, Buford J, Gurbani V (eds) Advances in next generation services and service architectures. River Publisher, Aalborg, 43–63
Moiso C, Minerva R (2012) Towards a user-centric personal data ecosystem the role of the bank of individuals' data. In: 16th international conference on intelligence in next generation networks (ICIN). IEEE, Berlin. pp 202–209
Neumeyer L, Robbins B, Nair A, Kesari A (2010) S4: distributed stream computing platform. In: IEEE international conference on data mining workshops (ICDMW). IEEE, Sydney, Australia, pp 170–177
Nielsen J (1998) Nielsen's law of internet bandwidth. Report on line, Useit
Ohtsuki H, Hauert C, Lieberman E, Nowak AM (2006) A simple rule for the evolution of cooperation on graphs and social networks. Nature (Nature) 441(7092):502–505
Rappa MA (2004) The utility business model and the future of computing services. IBM Syst J (IBM) 43(1):32–42
Rhea S et al (2005) OpenDHT: a public DHT service and its uses. ACM SIGCOMM computer communication review. ACM, New York, pp 73–84
Rosenberg J et al (2002) SIP: session initiation protocol. RFC 3261, internet engineering task force

Schulzrinne H, Rosenberg J (1998) A comparison of SIP and H. 323 for internet telephony. Report, pp 162–167

Shafiq MZ, Ji L, Liu AX, Pang J, Wang J (2012) A first look at cellular machine-to-machine traffic: large scale measurement and characterization. In: Proceedings of the 12th ACM SIGMETRICS/PERFORMANCE joint international conference on measurement and modeling of computer systems. ACM, London, UK, pp 65–76

Singh K, Henning S (2005) Peer-to-peer internet telephony using SIP. In: Proceedings of the international workshop on Network and operating systems support for digital audio and video. ACM, Skamania, Washington, USA, pp 63–68

Skype Wikipedia (2013) Skype. Wikipedia. http://en.wikipedia.org/wiki/Skype. Accessed May, Last accessed 27th, 2013

Tarkoma S (2012) Clean-slate datacentric pub/sub networking. Wiley Online Library, Sebastopol

Thom GA (1996) H. 323: the multimedia communications standard for local area networks. IEEE Commun Mag (IEEE) 34(12):52–56

Turaga D et al (2010) Design principles for developing stream processing applications. Softw: Pract Exp 40(12):1073–1104

Upton E, Halfacree G (2012) Meet the raspberry Pi. Wiley, New York

Vemuri K (2000) Sphinx: a study in convergent telephony. In: Proceedings of the IP telecom services workshop 2000 (IPTS2000). ATT Lab Research & Pulver.com, Atlanta, Georgia, USA, pp 9–18

Weiser M, Gold R, Brown JS (1999) The origins of ubiquitous computing research at PARC in the late 1980s. IBM Syst J (IBM) 38(4):693–696

Werner MJ (2013) Peer-to-peer networking using open web technologies. Research report. Faculty of Engineering and Computer Science of Hamburg University of Applied Sciences, Hamburg, Germany

World Economic Forum (2011) Personal data: the emergence of a new asset class. White Paper. World Economic Forum, Cologny

World Economic Forum (2013) Unlocking the value of personal data: from collection to usage. White Paper. WEF, Colgny

Chapter 6
TelCos Approach to Services

Telcos have been in charge of network access, control, and intelligence for many decades, but the service 'webification' has left them unable to compete in the service arena. Now, Operators seeking to transform the service layer, as well as the Voice session control, find that IMS will only help to migrate traditional services. The IMS framework does not offer a web-like application layer, but only a SIP-based interface to application servers. While it is suitable for large Voice systems, the session controller restricts the logic flow, unlike web applications that are free to initiate and integrate other services. Even the combination of IMS and SDP is not agile enough to give operators the competitive edge that they need.

Operators should explore other ways of exploiting their network assets. There can be several options of collaboration with application providers if open APIs are exposed and network facilities are made available. New opportunities are brought about by softwarization and virtualization, enabling the network to provide access to repositories of knowledge and to a variety of network resources.

6.1 Introduction

Many Operators are trying to enforce the usual business models and the traditional approach to services and infrastructure control. This is reflected, for example, in the evolution of the specification of 3GPP for the Long Term Evolution, LTE, and the following releases. From the service perspective, the major novelty of LTE is in the full switch to packet switching and the lay down of circuit switching, while simpler technologies for radio access are enabler for supporting the broadband IP connectivity. At the service and control level, LTE is not introducing relevant differences with respect to IMS. In the fixed network side, a similar approach is followed in ETSI TISPAN (TISPAN 2009). This can be seen as the willingness to reuse and

R. Minerva and N. Crespi, *Networks and New Services: A Complete Story*,
Internet of Things, DOI 10.1007/978-3-319-33995-5_6

capitalized the current investment in the network control and service infrastructure. However this attitude, especially in a period of rapid technological transformation, is inappropriate. The technological basis of these architectures are very old (more than ten years old) and new technologies and solutions have emerged and seized large sectors of the ICT market. Hadoop and the NoSQL technologies can be examples, as well as different programming languages and paradigms (agile programming). Pursuing with this old perspective gives a technological advantage to the Web Companies and other competitors. Instead of running after the Webcos technologies, Telcos should leap forward metaphorically and experiment, develop, and use alternative technologies.

In this chapter, an analysis of the most consolidated and used technologies "in the network" is provided in order to understand whether they are fit for supporting the Telcos in the daunting task of competing with innovative and technology oriented Webcos. First, an analysis of **current network platforms** based on the Network Intelligence is carried out. These platforms should allow for **easy development of services** capable of competing in richness of functions and easiness of usage with those offered by WebCos or by P2P systems. The **interaction and control paradigms** play once more an important role. In addition, the strong tie of the network and the services has to be evaluated and it should be understood if this is a real asset or a weakness in the development of services. The association of IMS and Service Delivery Platform, SDP, and the progressive move toward Service Oriented Architecture, SOA, within a Telco environment must be evaluated from the perspective of new classes of services.

A second scrutiny is devoted to the Exposure, i.e., the possibility to open up the Telco's systems in terms of **Application Programming Interfaces** that can be freely used by third parties in order to develop networked applications. Are they a real business enabler? Is there a practical usage of network-related APIs? What kind of interfaces should be exposed? Actually, there is a property of the APIs that needs to be considered: the more the API is abstract from the underlying resources, the simpler is programming, but there is also a deficiency in expressive power (i.e., the API provides too simple and sometimes useless functionalities).

A third issue to consider is the assumption that users want Quality of Service, **QoS**, and Quality of Experience, QoE, from the network. This is a recurring argument and it is profoundly determining the strategies and the attempts of many Operators to be successful in the services realm. Is there any perceived value from the user stand point in network services offering some guarantees in the Quality of Service? Is it an appealing form the **residential users**? Are these features appealing to the business market?

The underlying issue analyzed in this Chapter is to determine **whether the traditional properties of the "network" are still assets** or are simply an attitude that the Telcos should get rid of.

6.2 The Network Intelligence Legacy

The IP Multimedia Subsystem (Copeland 2009) (first defined around 1999) is a specification for an IP mobile network, originally devised by the 3GPP as an evolution of the GPRS architecture (Bettstetter et al. 1999). The general goal of IMS was to provide a reference architecture for a new system which is able to encompass voice and multimedia services in an all-IP infrastructure. In particular its goals were the following:

- To decouple the telecoms infrastructure into different layers (namely Transport, Control, and Service Layers) in order to guarantee the evolution and independence of networked functionalities;
- To provide voice and multimedia services over a controlled (mobile) IP network;
- To support different access networks (e.g., UMTS but also WLAN, WiMax, etc.);
- To align with the Internet and possibly to reuse existing IETF protocols to get a full compatibility with the IP and Internet standards.

The definition of IMS first caught the attention of Fixed Line Operators. At that time, fixed network issues were focusing on the definition of a viable architecture able to integrate the circuit and the packet switched infrastructures. Many Telcos still want to provide multitude of traditional services that they were contracted to continue maintaining. The fixed line community preferred the IMS standards over the Softswitch. The emergence of the IMS architecture helped resolving these questions, so the fixed line community embraced the IMS concepts. At the same time, ETSI TISPAN was working on the definition of IMS architecture extensions that support Fixed–Mobile Convergence.

Figure 6.1 depicts the IMS architecture now accepted also by ETSI TISPAN as a framework for Fixed–Mobile Convergence.

The IMS architecture is based upon the separation of functionalities at transport, control, and service levels. The main protocols are IETF protocols (with some changes): SIP (Rosenberg et al. 2002; Johnston 2009) for the control of sessions of services and Diameter (Calhoun et al. 2003) (for management and QoS related operations).

The transport layer is derived from the GPRS architecture. The Mobile Access Network collects IP streams and forwards them toward a couple of nodes, i.e., the Serving and Gateway GPRS Support Nodes (SGSN and GGSN) that guarantee the transport of the IP packets from the access network to the Internet. The IMS transport layer has been extended in order to allow for the integration of fixed access (through Digital Subscriber Line Access Multiplexer (DSLAMs) and Border Gateways) and WLAN networks. These forward IP packets directly into the core IP network. In order to provide some functions related to QoS, the Diameter protocol and its control architecture have been adopted. There are Policy Decision Points (logically pertaining to the control infrastructure), that decide which policies to

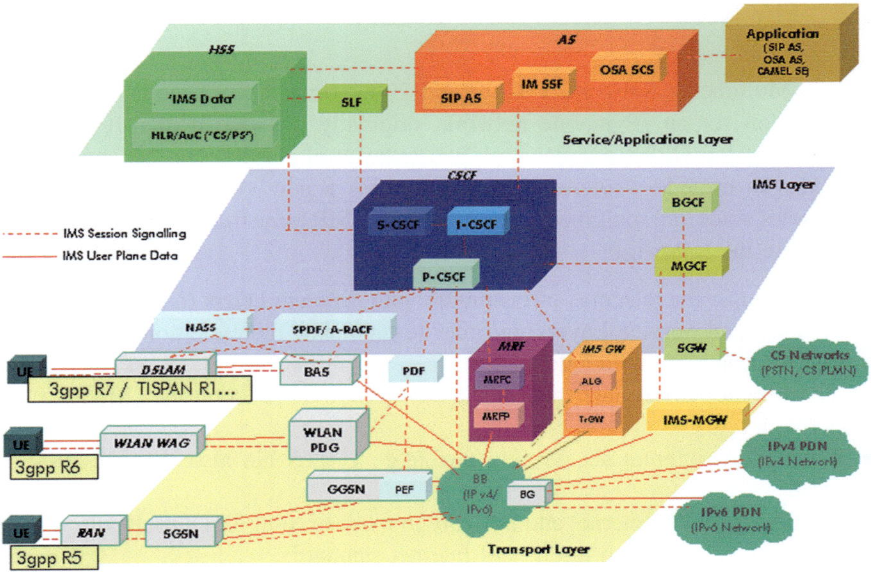

Fig. 6.1 IP Multimedia Subsystem

apply for a certain stream of IP packets) and Policy Enforcement Points that execute those policies. In order to interact with existing PSTN (Public Switched Telephone Network) networks a media gateway function is defined, it allows for the "translation" of VoIP streams into analog voice. In addition, Media Resource Functions are defined as a means to provide multimedia-related functions.

The Control Layer is organized around a number of servers. IMS components essentially control session management in the Call Session Control Function (CSCF), and the interworking with the PSTN signaling system. The CSCF Servers are duplicated in order to allow for roaming and mobility. This is essentially the equivalent of the call control mechanism implemented in the PSTN. In addition, the IMS standards include the evolution of LHR to HSS (Home Subscriber Server) and the policy server (PCRF) that determines network QoS. The HSS supports the service layer by combining several AAA functions for the management of user data and identity as well as mobile network location function application servers. The Service Layer points to a loose integration of SIP, OSA, and IN (CAMEL) functional entities. They are able to communicate with the control layer by means of the IMS Service Control interface (based on SIP), ISC. At this layer, Services exposure gateway enables third party application providers to use functions and services of the IMS infrastructure.

This architecture has been promoted as a step forward in the field of multimedia communication by the Telecommunication industry, and was hyped beyond its purpose and capabilities. It has received a number of criticisms (e.g., Waclawsky 2005; Manzalini et al. 2009; Minerva 2008). In fact, the ability of the IMS architecture to provide new services has been questioned, since it mirrored the Telecom approach, while services were clearly provided by the web environment. The IMS is defined on the assumption that services' session control will be "centralized" and controlled by the Operator (as in the old Intelligent Network architecture). This approach tends to reapply existing business models and solutions in the new context of Next Generation Networks, where established Operators and Vendors push for a "controlled" transition to a more Internet-based control of services but still defending their legacy networks and product lines. IMS proposes the usage of only two protocols (SIP and Diameter) in order to provide session related capabilities. SIP seems to be a new signaling system, but the traditional framework for providing services remains the same, in the effort to support the large installed base of IN and Camel applications.

In IMS, the session concept is synonymous of "call," in fact services are built on top of the concept of establishing a session. The session is "stateful," i.e., the states are stored in the network in order to keep track of the progress of processing in terms of finite-state machine. CSCF is essentially an evolution of the Basic Call Model by means of the SIP functionalities. In addition, the IMS architecture (based on the successful GPRS one) is introducing the traditional mechanisms for the (artificial) perimeterization of services within the domain of an Operator. In this sense, the CSCF functionalities are replicated in order to move the control to "visited" Operators that have contract with roaming customers. This framework was complex due to the break down of functionalities into discrete servers. As a result, many inter-server SIP interactions take place in order to execute control functionality, and even higher if roaming functionalities are needed.

From a service perspective, the greatest drawback comes from the lack of definition in IMS of the Service Layer. It is generically referenced as an environment in which different applications servers coexist (SIP, OSA, and Camel), where they use different protocols, but need to interface to IMS via SIP. This means that a lot of "normalization" has to take place.

Another barrier to adopting IMS was the management protocol—Diameter. It replaced Radius, which was deemed unsuitable, but had already wide installed base. Since IMS only provided Diameter as the management protocols, Operators had to give up from using COPS (Common Open Policy Service Protocol) to enforce QoS and Policy.

The main issue with IMS is that the interaction paradigm between the session control and the service layer is still the traditional one of Intelligent Networks, i.e., services are triggered by network requests from the terminals—not the applications. The session control recognizes triggers for services and passes control to the application servers, but when the service logic has been executed, the control is returned to the Control Layer. Furthermore, chaining services is difficult, because the session control is involved in every hop, although services could call other services before returning control to the CSCF. This is much more constrained than the freedom and the expressiveness capabilities offered by the Internet Application Servers, which are using much easier interfaces (e.g., based on the REST architecture).

In order to overcome this major shortcoming of the IMS architecture, many Operators have found a solution in a better definition of the Services Layer, in particular the concept of Service Delivery Platform, SDP has been advocated as a solution to a missing definition of the service environment within IMS.

SDPs (Ohnishi et al. 2007; Moriana Group 2004), are a way to create and govern services over an IP Platform. SDPs (see Fig. 6.2) are telecom-oriented platforms that use SOA (Sward and Boleng 2012), and Web Services (Snell et al. 2009) technologies in order to enable a framework for building new services. The concept spawns from the application server definition: they embed within the IT framework "connectors" for interfacing with telecom networks and telecom devices. The idea is to be able to bridge two classes of servers: the telecom control servers (such as Parlay Gateway) and the IT application servers. The goal is to create a sophisticated environment to develop, manage, and deliver services, as well as reducing the development and deployment costs associated with telecom services.

An SDP is built around a mechanism for passing and controlling events between a number of processes, each one representing a reusable functionality. Such a "service bus" allows communication and synchronization (by means of exchange of messages, i.e., events and commands) between cooperating processes and functionalities. The "bus" allows the flow of some messages also toward external domains, so enabling the communication with third parties applications. Functionalities (and related processes) are divided into: Enablers, that provide common and general purpose functionalities; and Services, that use the general functionalities to build business logics and services. Some Services and even some Enablers (through the exposure interfaces) can be provided by third parties. In this way, the service portfolio and the functionalities offered by an SDP can be greatly extended.

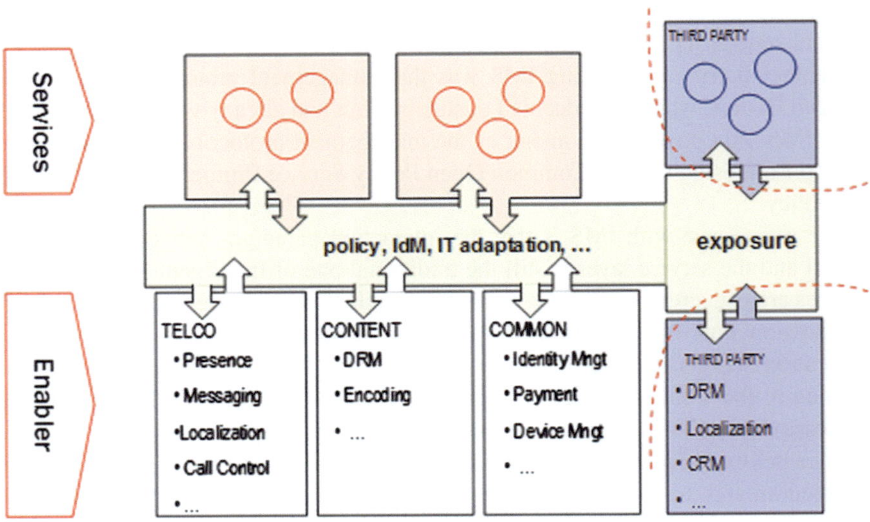

Fig. 6.2 General structure of a Service Delivery Platform

Telecom systems often combine SDP with IMS. The reason is that IMS did not define a common service architecture (i.e., a software architecture is needed and not a functional one). In other words, the IMS is a collection of modules that are used to provide session control (location, identity authentication, QoS, and policy), but not a service architecture that provides common service logic modules, storage and data management, etc. SDP seems to cover (in a traditional operators' approach) the lack of specification for a well-structured and well-defined Service Layer.

The combination of IMS and SDP is viewed by many operators as a winning combination to provide a rich set of session and content-based services over the new all-IP network. In addition, the SDP approach is a means to leverage the possibility for Operators to offer Network APIs that could appeal developers and foster the creation of a service ecosystem centered around the assets and the traditional approach of the Operator. As deeply discussed in Manzalini et al. (2009), this approach is not working because, simply put, the real development platform is the Web. It is in the web context and with the web protocols and mechanisms that services will grow. In addition, as discussed in paper on paradigms, the programming paradigm chosen for the IMS and SDP combination are very far from the predominant ones of the Internet (essentially based on a client—server model and not on triggering mechanisms stemming from a complex finite state machine as proposed by IMS). To make things even worst for the Telcos, the technological gap between the Web Companies platforms and the IMS-based one is increasing and not reducing because the developer's communities are bringing a lot of new ideas and initiatives. The success of Telcos ecosystems based on NetAPIs and SDP is very limited even in the consolidated realm of Voice Services (e.g., the integration of Rabbit Ribbit in BT).

The SDP promise was also aiming at the integration of cloud and XaaS approaches into the realm of the Telcos. As discussed in Minerva et al. (2013b), Telcos have limited possibilities also in the cloud market. If they want to success, they have to aim at the business market and specifically to those companies that want to integrate their existing processing and storage infrastructure at an international level and have a certain level of QoS with respect to interconnection between systems distributed in several Countries. The Federated Cloud approach is the most promising one for the Operators because it allows to leverage several assets as: local presence, continuous support to the customer at a national level, some level of guaranteed communication services, the need and the possibility to interwork and integrate different clouds. This is a small part of the entire cloud business that gives an advantage to the Web Companies if customers are looking for cheaper price, simplicity of usage, and greater availability of resources.

In order for the Operators to be competitive, it is necessary to disrupt the market with novel service propositions and at the technological level by introducing discontinuity in the status quo. The SDP + IMS approach is a mainstream approach that points to give a programmable control on traditional service but does not support innovative service deployment. IMS in particular is a traditional platform that suits only the traditional value-added service approach. Being a horizontal platform, it may serve for providing also other classes of service, but it lacks specialized and more IT and web-oriented features. An example is the Video Services provision that can be implemented with SDP + IMS, but the rigidity of this platform makes it less adequate than other specific platforms for the provision of TV-like service.

Balboni and Minerva (2012) discusses how a Telecom Operator can introduce new services and what are the new technological paradigms that could be followed. One direction is aiming at user-centric services such as Personal Data, Identity Management, and Security adopting a disruptive approach: helping the user to better cope with an unfriendly Internet interested in spoiling the user's rights. The second direction is to move from Internet of Things (IoT) to the Internet with Things, i.e., the environment capable of representing each real-world object and to allow the creation of many applications and the servitization of products. The third direction is toward the smart environment and the edge intelligence, i.e., the possibility for the Operator to act as a sort of anchor point (a hub) around which edge node can aggregate in order to create dynamic and very smart environments. In order to take these directions, there is a need to rethink the traditional Operator approach and to develop new platforms that disrupt the current technological mainstream. In this sense, the new platform can be based on overlaying (especially using P2P technologies), on cognitive behavior of nodes and the entire infrastructure, on virtualization of resources and functions. This is a new combination of existing technologies that can help in competing with the WebCos. In Minerva et al. (2011, 2013a), a detailed description of the envisage architecture can be found. Here the most important properties of it are briefly sketched out. A new Service Enabler Platform (a network of networks) will strongly depend on the following:

- the virtualization of resources (communication, processing, storage, sensing);
- the cooperative orchestration of single or subsystems' resources;
- the introduction of self-organization capabilities in order to achieve an autonomic, cognitive (Manzalini et al. 2012) behavior of applications and resources (the highly dynamic and unpredictable behavior of a network of networks requires the real-time adaptation to different contexts);
- the introduction of different paradigms for the cooperation of distributed (virtual) objects. Overlay networking is a real distributed processing paradigm and it fits properly in this dynamic environment. The combination of these technologies will lead to a programmable networked environment such as the one represented in Fig. 6.3.

The Service Enabler platform is a sort of Network Operating System that through the representation and virtualization of networked resources spanning across many subsystems and different administrative domains, will allow applications to negotiate for "virtualized" and autonomic resources, to allocate them, to control and program their functionalities according to the specific needs and requirements. The upper layer is made out of overlay network that comprises basic resources. These basic resources can be extended or can be integrated with new

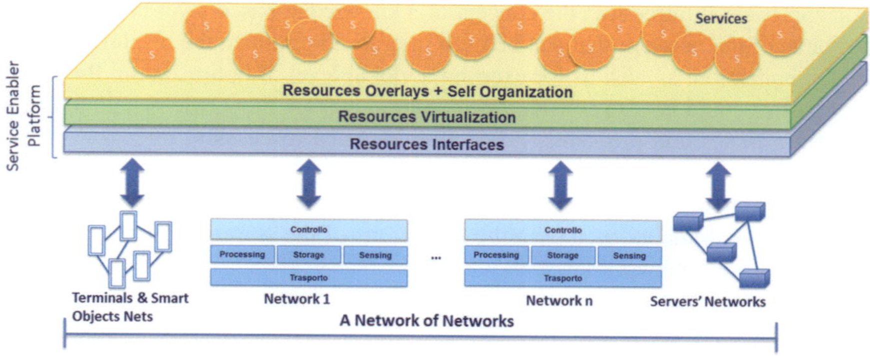

Fig. 6.3 A service enabling environment and its basic principles and layering

specialized ones in order to allow for the provision and offering of many services. It is important to stress out the role of end-users terminals and networks. They provide to the entire network a set of capabilities and the possibility to the entire network to rapidly grow (similarly to P2P networks in which end users contribute to the resources of the system that can scale up).

Actually virtualization and opening up of interfaces (Application Programming Interfaces, APIs) can be combined in order to build a very flexible and programmable platform for services. Figure 6.4 depicts this possibility from a clean

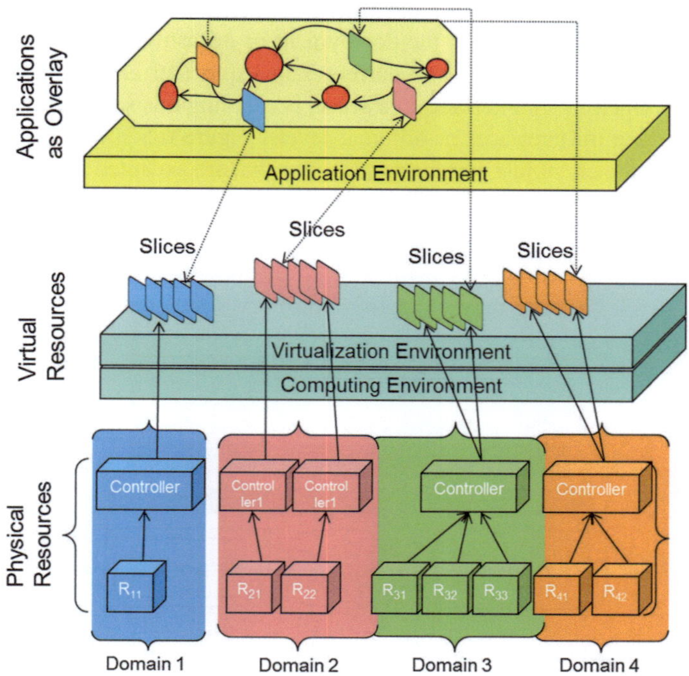

Fig. 6.4 Layered cognitive architecture for the future Internet

slate approach. It introduces the concept of a knowledge-based layer that represents the status of the network that can be used in order to optimize the resources for the single application, as well as for exerting a global level of optimization. Predictive mechanisms could be also considered in order to prepare the whole system to tackle upcoming issues and problems. For a more detailed discussion, refer to Minerva et al. (2011).

One important feature of this platform is its capability to span over different administrative domains exceeding the limitations of the current strong coupling of service layer and network layer.

6.3 About Network APIs

Many Telcos see the possibility to monetize the network by means of APIs and related functionalities. Virtualization and the progressive softwarization of network functions lead to new issues and new opportunities for positioning the APIs in the market. One major point related to softwarization is its huge impact on the value of the network. Up to now, network deployment is a capital-intensive business, i.e., creating a new network means to put upfront a relevant amount of money and to have a return of investment in the long term (actually many business plans of Operators have shown that the payback period for new deployment in LTE and Fiber is increasing from 4–5 years to over 10 years and usually it scale up to 15 years). The softwarization of the network can break this barrier and many more Actors can be ready to invest in the deployment of infrastructure. Many Operators see this as a direct menace to the current status quo and an excessive opening competition. Opening up interface, in a sense, is enabling this scale down of barrier. Telcos can open up interfaces at different levels. Figure 6.5 shows the most conservative one: the client just pay for applications that are executed in the framework of the Operator.

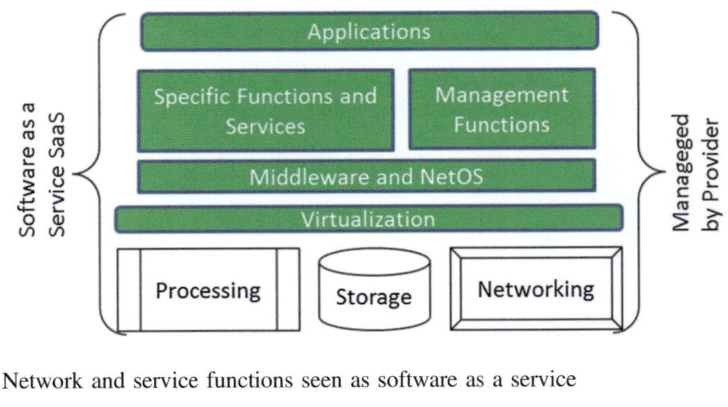

Fig. 6.5 Network and service functions seen as software as a service

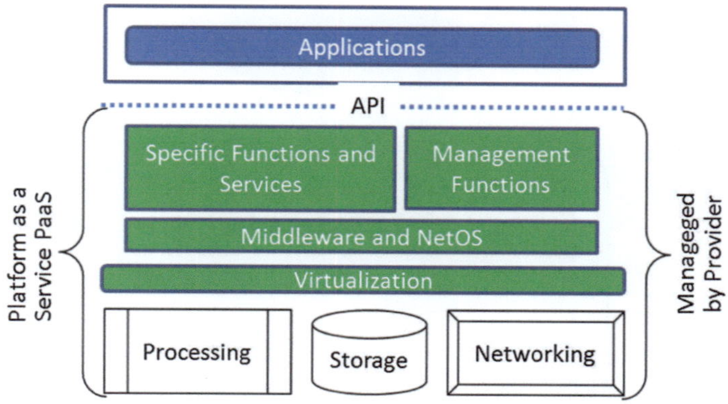

Fig. 6.6 Network as a Platform (PaaS)

In this case, the clients use the services offered by the network as such, the capability to modify the functions are to be negotiated directly with the Provider, while the direct access to resources and low-level functionalities is not offered and the "ownership" of the infrastructure is kept in the hands of Operators. This represents the situation in which a Telco sees its infrastructure and service offering as a sort of Walled Garden totally closed and controlled by the Operator. Another approach is to position the infrastructure as a platform that offers value functionalities and open up APIs for programming them. Figure 6.6 depicts this possibility.

In this case, the programmable part of the network comprises value-added functions which are specifically networking, storage, or processing functions that the applications can use and orchestrate in order to reach their computational and communications goals. One example of this possibility is in the context of content delivery networks (or information centric networks): a web company can access and use these functionalities in order to build an abstract CDN according to specific needs. The Operator does not actually disclose the network topology, but it offers the possibility to create a virtual CDN according to a number of local and geographical requirements. The advantage is a higher level of programmability and abstraction for the Client and, for the Operator, a major grip on the actual network infrastructure (that is not disclosed at a physical level).

Figure 6.7 represents the need to have different APIs in order to satisfy different requirements from potential customers.

Figure 6.7 shows that three different interfaces can be offered: the North interface is following the traditional approach of the Operators, the network infrastructure triggers events and the application running in the cloud can use this interaction paradigm in order to execute services and applications. The East interface is offered to Providers according to Web technologies and interfaces. Providers can use this interface in order to organize an abstract network according to their needs. The well-known web interactions (client–server relations) will be used in order to regulate the usage of value added functions. In this case, the

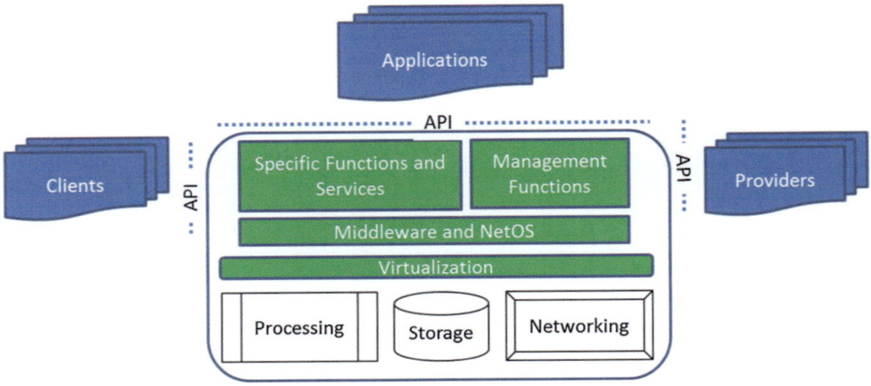

Fig. 6.7 North, East, and West bound interfaces for a Network Platform

network functions are seen as services and functions offered by servers, i.e., the providers are "the clients" in a client–server relationship. The West Interface is devoted to final users that can use it in order to adapt the network to its specific needs. Examples of this usage fall in the realm of changing dynamically the characteristics of the access to the network (varying some parameters, e.g., a gamer can dynamically request to the network to change the value of maximum latency). The real commercial value of this interface is questionable, however, some P2P applications or communities can use this interface in order to better serve the specific community requirement. Also in this case, the API should be exposed making usage of web and/or P2P interfaces and technologies. This arrangement of interfaces preserve to the Operators the entire value of the infrastructure (made out of processing, storage, and communications capabilities).

Figure 6.8 shows a more granular and programmable approach. Here the resources are virtualized and they expose direct control interfaces.

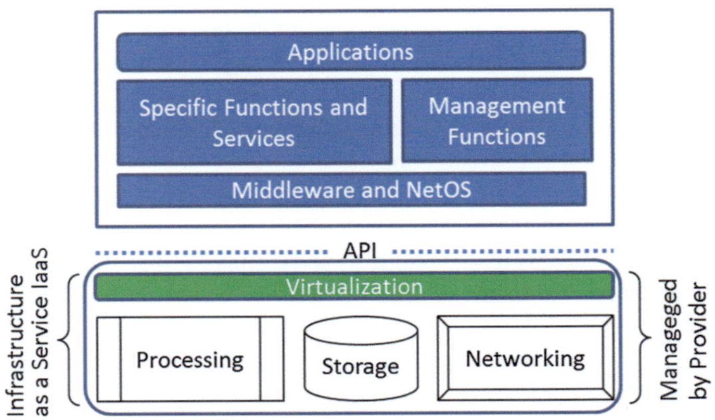

Fig. 6.8 "Network as a Service"

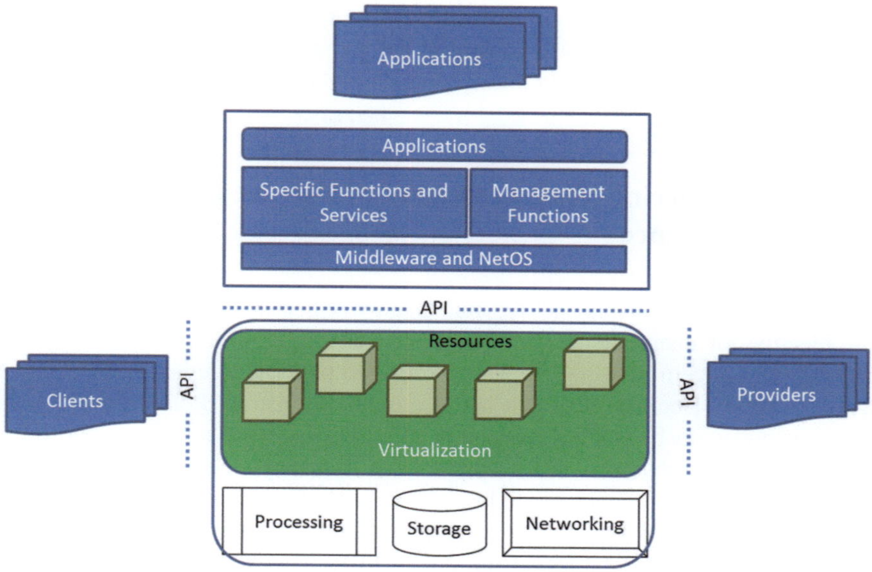

Fig. 6.9 Network resources as a service

In this case, the programmers have direct access to virtualized instances of the network resources. They can program and orchestrate the resources in order to create their own virtual infrastructure. The control on each allocated virtual resource is granular. This means that the Network functions are totally spoiled and the control is moved to the programmability infrastructure. In terms of APIs, the possibilities are similar to those discussed for the Network as a Platform case. Figure 6.9 shows the three identified APIs.

In this case, the North bound interface is defined and used in a "network intelligence" based paradigm, i.e., events and commands are passed though APIs that are capable to deal with this interaction model. Applications and middle level functionalities can be placed in a cloud (eventually hosted by the Operator itself). In this case, the control is very granular and the programmers have full control of the allocated resources. The East interface is offered to Providers that can use this function in order to allocate and control specific resources and to orchestrate them for the benefit of the global goals of the Provider. This interface is directly determined by the type of resource and eventually it could be offered according to the preferred paradigm (event based, Client–Server, or Peer-to-Peer). The West interface (and also in this case its commercial use is questionable) should offer similar mechanisms of the East bound one. Its envisioned use is for prosumers or clients that want to favor the development of P2P or community applications. APIs at this level provide a granular control of each virtualized resource and they give a great deal of "ownership" to the users. The effect of this is lowering of the importance of owning the physical resources and the reproducibility of resources. As said, this can break the barrier of intensive capital investment on the network. The combination of softwarization and the generalized usage of general purpose computing can lead to many new entrants to create alternative network solutions or part of networks.

6.4 The Recurring Myth of Quality of Service and Quality of Experience

The provision of negotiable and dynamic Quality of Service for All-IP Networks is an objective for many architectural definitions. IMS has been defined around the possibility to govern network elements in order to provide differentiated classes of services. Policy Decision Function nodes are able to instruct Policy Enforcement Function nodes on bandwidth management, class of service allocation and other QoS-related characteristics. This approach is typically "network intelligence" based, i.e., a small number of nodes (at the control and service level) provide intelligent functions in order to govern several underlying simple network elements. Nodes at the upper layers decide the policy to be executed, while the transport elements enforce the chosen policies. This is a sort of Event–Command paradigm in which the network elements are capable to detect triggering conditions and to send events to the upper layers. Nodes at this level are decision points that determine the logic to be executed and instruct the enforcement points on how to proceed by means of commands (commands can also be predefined scripts created using a QoS definition language). This control model is not aligned with the basic concepts of IP communication, e.g., the introduction of a supervising layer of governing nodes is not aligned with the independence of autonomous systems. This approach is difficult to introduce for design, technical, and economical reasons and it could lead to a misalignment in the industry: on one side, the Operators will force a modification in IP networking, and the networking community will enforce the best effort principle. The best effort mode has allowed the creation of large networks made out of cooperating autonomous systems and the creation of many applications that (using asynchronous functionalities) are able to provide an acceptable level of end-to-end quality of experience. The cooperation of Autonomous Systems (AS) is the key advantage of IP networking, different networks (often operated and controlled by different entities) cooperate in order to find a viable path for routing packets. The shared goal of AS is to maximize the number of packets that can be effectively routed toward a destination (either an end point or another network). This leads to two considerations:

- Users perceive quality of service as the result of cooperation between independent systems (Fig. 6.10);
- Enforcing quality of service characteristics into a single network does not guarantee the enforcement of the quality in all the chain of cooperating subsystems.

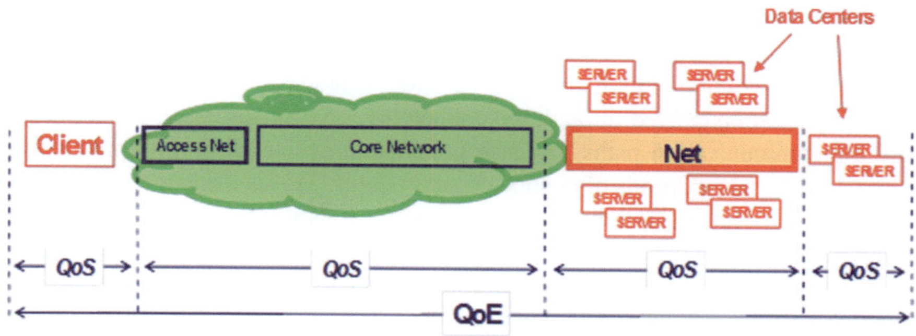

Fig. 6.10 Quality of experience as perceived by the user

In the future, these subsystems will not only provide communication, but also processing, storage, and "sensing" capabilities. This transformation will exacerbate the problem of governing control and management states of resources allocated to AS. There are many management dependencies among resources. These relationships are not fully updated in an automatic manner. Increasing the number of resources will create an even greater number of dependencies and the need to automatically update them will clearly emerge. Moreover, high-level network goals (e.g., connectivity matrix, load-balancing, traffic engineering goals, survivability requirements, etc.) are translated into low-level configuration commands individually executed on the network elements (e.g., forwarding table, packet filters, link-scheduling weights, and queue-management parameters, as well as tunnels and NAT mappings) which are mostly handmade. For example, IP and Ethernet originally embedded decision logic for path-computation in distributed protocols, whose complexity incrementally grew with the growth of Internet. Growing complexity and dynamicity in future networks pose serious doubts about the efficiency in extending further the distributed control protocols. On one side, the network decision logic should become scalable and more effective. On the other side, low-level configurations should be made adopting self-configuration capabilities (triggered by global network decision logic).

Quality of Service and different Networks

The problem of guaranteeing a system-wide Quality of Experience depends on the ability of the entire communication and computation environment to support the requested features and constraints. As seen in Fig. 6.10, this capability has impact on each subsystem (from terminals to public and private networks): each component should be able to support the expected level of quality, otherwise the entire environment will be providing a quality of experience depending from the lowest quality provided by the less capable subsystem. This degradation effect is not the only one to cope with: even if all the involved subsystems were individually capable of providing the expected quality of service, this does not guarantee the final result because each subsystem has to coordinate and cooperate with other "neighbors" in order to achieve the expected result. Except in environments totally under the control of a single provider, the QoE is an issue that has to be tackled with a different perspective aiming at the cooperation of independent systems. The QoE problem should be tackled through the cooperation of ASs aiming at providing not only the communication paths, but also at cooperating for supporting processing, storage and "sensing" needs of connected subsystems. This is a daunting task, but it is one in-line with the Internet approach and it is also encompassing needs and requirements of networked computing (grid and cloud computing) as well as those of the Internet of things. A federated approach in which a subsystem is able to provide a set of functionalities to others and its cooperation for achieving a common goal should be the basis for approaching the QoE problem. There is another issue related to QoE in this large distributed systems: complexity. A network of networks tends to collect a number of heterogeneous systems and nodes that create the need for very complex integration and management. With the increase in number of nodes, the traditional management of large system will fail short, so there is a need to determine how the components of a network of network should be organized, configured, and managed in order to provide better QoE. Actually virtualization and segmentation of functionalities can allow the Provider to request the Network Provider to allocate virtual resources that support certain level of expectations. These resources can span over different administrative domains and can interoperate in order to provide the required level of service (even following a best effort approach, in fact the single environment will adhere to the current way of providing services, but the virtual infrastructure could be over equipped in terms of (virtual) resources so that the negotiated QoS parameters can be meet (at least statistically).

6.5 Conclusions

This section has examined some of the core beliefs of many Telcos. Network Intelligence, Exposure and QoS are persistent icons of the past. Some Telcos believe that they are still key differentiators with respect to new services. In this perspective, the important thing is to show that their proper usage can return a lot of value to Operators allowing them to recover importance (and revenues) in the market. This Chapter, however, has shown that the current platforms are obsolete (their definition is more than ten years old) and there are new technologies (such as, virtualization and software defined networking) that can provide better programmability and can support the deperimeterization of services (from the network). The IMS has been conceived as a novel architecture for providing session-based services. For session-based services, the old traditional telephony services are mainly considered. IMS can fully support them, however newer architectures like P2P have clearly demonstrated that other technologies can now be more efficient and less costly in order to support these services. IMS has been also proposed for multimedia services (Jain and Prokopi 2008). Surely the platform is capable of sealing with them, but the market is much more oriented toward specialized solutions devoted to IPTV and multimedia services in general. IMS has also been considered as a viable platform for supporting other services (like the M2M ones). Also in this case, newer and more advanced solutions can be used. For instance, Twitter could be transformed into an efficient enough engine for many IoT applications. The control paradigm based on a network capable of triggering events toward the services layer is adaptable to many situations and services, but the fact that services are depending on connectivity and related basic call model is too much of a burden. NetAPIs or Exposure is another Telco activity that started too soon and never found a convincing path to the market. Many Telcos were ready in early 2000 (just when the new Web 2.0 trend was taking shape) to release Parlay-based APIs. What stopped a large development of them? Essentially three considerations emerged: why to offer network control functions if nobody was pressing for having them? What is the right charge for these functionalities? What are the security risks? These three considerations stopped almost any initiative in this area. It is singular that after more than ten years, many Operators are re-proposing very similar approaches even if it is evident that two of the major attempts in this area (rebbit and bluvia) have failed. Proposing this approach now is not a winning proposition: even less people are interested in call or session-based control (and if they are, they find appealing interfaces somewhere else), There is a lack of standardization and APIs are fragmented. Initiatives like oneAPI of GSMA are lagging behind and they are not delivering interesting enough APIs for developers to use. Finally, the Quality of Service issue is a driving concept within many Telcos. There is still the opinion that QoS is a property requested by the market that differentiates the Telcos from other providers. It is difficult to change this attitude. And many Telcos do not want to learn lessons from the past. For instance, when the Google maps were first released, users were experiencing very long waiting times before the maps were downloaded. And when a user was scrolling too fast over a part of the map, there was a long delay before the new part of the map was shown. At the time this was taken as evidence that the Internet needed to have more bandwidth and it should be also supported by QoS properties. In other terms, the Operators were saying that in order to solve the problem, the network should provide more capacity and this capacity should be governed by means of signaling and policing. The same problem has been tackled and solved by the web in a totally different way: the browser should asynchronously request the most probable requested pieces of the map before an explicit request of the user. Simply put the browser is downloading more data than actually shown to the user and they are used as buffers while other data are downloaded in order to fulfill a user request. This asynchronous request was the basis for a winning web technology: AJAX. This example shows that the attitudes of Webcos and Telcos are deeply different and one is promoting solutions at the edge, while the other is enforcing policing and capacity in the network. With the increasing capacity offered by the new networks and with the growing capabilities of terminals and servers, there is a possibility (as demonstrated by the Amazon's Silk solution) that the cooperation between the terminal and the cloud can solve many communication issues. Actually this seems to be a very promising possibility, depending on the available network resources, processing, and flow of data can move toward the terminals (if the conditions allow this) or toward the cloud that will take care of optimizing the delivery of information to the terminal. The real progress in distributed applications and for new classes of services relies on the fact that the network resources have to be programmable and they should be virtualized. By means of these two properties, virtual networks tailored for specific needs of applications can be created and programmed. For instance, a Content Delivery Network can be virtualized and can be deployed over an SDN-based infrastructure. Nodes become programmable and they can be migrated also toward processing infrastructures not related to the Operator. In this way, deperimeterization of the services from the network can be pursued. As a conclusion, it could be stated that neither the NI Platforms, nor the API Exposure and nor the QoS platforms are mechanisms capable of solving the issues that Operators are facing nowadays.

If Telcos want to play a different role from access communication provider (ad bit pipe), they need to use the technologies discussed in the book in order to reinvent themselves from a technological perspective. In addition, there is a need to change attitude also with respect to software development. Another associated issue is also to understand new business models and try to use the new technologies for implementing them. In other terms, the road to a new Telco passes through disruption in technologies achievement and disruption in business models.

Bibliography

Balboni GP, Minerva R (2012) Nuove Reti e nuovi servizi. In: Notiziario Tecnico Telecom Italia, pp 6–17

Bettstetter C, Vogel H-J, Eberspacher J (1999) GSM phase 2+ general packet radio service GPRS: architecture, protocols, and air interface. IEEE Commun Surv Tutorials (IEEE) 2(3):2–14

Calhoun P, Loughney J, Guttman E, Zorn G, Arkko J (2003) Diameter base protocol. RFC 3588, IETF, Fremont California, USA

Copeland R (2009) Converging NGN wireline and mobile 3G networks with IMS. CRC Press, London

Jain M, Prokopi M (2008) The IMS 2.0 service architecture. In: The second international conference on next generation mobile applications, services and technologies, 2008. NGMAST'08. IEEE Computer, Cardiff, pp 3–9

Johnston AB (2009) SIP: understanding the session initiation protocol. Artech House Publishers, London

Manzalini A, Minerva R, Moiso C (2009) If the Web is the platform, then what is the SDP? In: 13th international conference on intelligence in next generation networks, ICIN, Bordeaux, France: ICEE, pp 1–6

Manzalini A, Minerva R, Moiso C (2010) Autonomic clouds of components for self-managed service ecosystems. Journal of Telecommunications Management, 3(2):164–180

Manzalini A, Minerva R, Moiso, C (2010) Exploiting P2P Solutions in Telecommunication Service Delivery Platforms. In N. Antonopoulos, G. Exarchakos, M. Li, & L. A., Handbook of Research on P2P and Grid Systems for Service-Oriented Computing: Models, Methodologies and Applications. Hershey, PA: Information Science Reference. pp 937–955

Manzalini A, Minerva R, Moiso C (2010) The Inner Circle: How to exploit autonomic overlays of virtual resources for creating service ecosystems. 14th International Conference on Intelligence in Next Generation Networks (ICIN) Bordeaux, France: IEEE pp 1–6

Manzalini A, Minerva R, Moiso C (2010) Towards resource-aware network of networks. 5th IEEE International Symposium on Wireless Pervasive Computing (ISWPC) Modena, Italy: IEEE pp 221–225

Manzalini A, Moiso C, Minerva R (2011) Towards 0-touch networks. Technical Symposium at ITU Telecom World (ITU WT). Geneva: IEEE pp 69–74

Manzalini A, Brgulja N, Moiso C, Minerva R (2012) Autonomic nature-inspired eco-systems. In: Gavrilova ML, Tan KC, & Phan C-V (eds) Transactions on Computational Science XV, vol 7050. Springer, Berlin Heidelberg, pp 158–191

Manzalini A, Minerva R, Moiso C (2012) Bio-Inspired Ecosystems for Telecommunications Pervasive Networks. In M. L. Howard, Pervasive Computing. Nova Publisher.

Manzalini A, Crespi N, Gonçalves V, Minerva R (2012) Towards halos networks ubiquitous networking and computing at the edge. 16th international conference on intelligence in next generation networks (ICIN) Berlin: IEEE pp 65–71

Manzalini A, Deussen PH, Nechifor S, Mamei M, Minerva RA, Salden A et al (2010) Self-optimized cognitive network of networks. Future network and mobile summit, Florence, Italy: IEEE pp 1–6

Manzalini A, Deussen PH, Nechifor S, Mamei M, Minerva RA, Salden A et al (2011) Self-optimized cognitive network of networks. The Comp J 54(2): 189–196

Minerva R (2008) On the importance of numbers and names for a 4G service architecture. In: Annual review of wireless communication, vol 3. IEC

Minerva R, Crespi N (2011) Unleashing the disruptive potential of user-controlled identity management. Technical Symposium at ITU Telecom World (ITU WT) Geneva: IEEE pp 1–6

Minerva R, Crespi N (2016) Networks and New Services: future of telecommunications. Springer

Minerva R, Manzalini A, Moiso C (2011) Which way to QoS in future networks: distributed or centralized decision logic? In: 50th FITCE conference—ICT bridging an ever shifting digital divide. IEEE, Palermo, pp 1–6

Minerva R, Manzalini A, Moiso C. (2011) Towards an expressive, adaptive and resource aware Network Platform. In A. Prasad, J. Buford, & V. Gurbani, Advances in Next Generation Services and Service Architectures. River Publisher, pp 43–63

Minerva R, Manzalini A, Moiso C. (2011) Which way to QoS in future networks: distributed or centralized decision logic? 50th FITCE Conference —ICT bridging an ever shifting digital divide. Palermo, Italy: IEEE pp 1–6

Minerva R, Manzalini A, Moiso C, Crespi N (2013) Virtualizing Network. In E. Bertin, N. Crespi, & T. Magedanz, Evolution of Telecommunication Services vol. 7768. Berlin Heidelberg, Springer pp 227–256

Minerva R, Moiso C, Manzalini A, Crespi N (2013) Virtualizing Platforms. In E. Bertin, N. Crespi, & T. Magedanz, Evolution of Telecommunication Services vol. 7768. Berlin Heidelberg, Springer pp 203–226

Minerva R, Manzalini A, Moiso C, Crespi N (2013a) Virtualizing network. In: Bertin E, Crespi N, Magedanz T (eds) Evolution of telecommunication services, vol 7768. Springer, Berlin, pp 227–256

Minerva R, Moiso C, Manzalini A, Crespi N (2013b) Virtualizing platforms. In: Bertin E, Crespi N, Magedanz T (eds) Evolution of telecommunication services, vol 7768. Springer, Berlin, pp 203–226

Moiso C, Carreras I, Fuentes B, Lozano J, Manzalini A, Minerva R, et al. (2010) Towards a service ecology for pervasive networked environments. Future Network and Mobile Summit. Florence, Italy: IEEE pp 1–8

Moiso C, Carreras I, Fuentes B, Lozano J, Manzalini A, Minerva R, et al. (2010) Towards a Service Ecology for Pervasive Networked Environments,. Future Network and Mobile Summit. Florence, Italy: IEEE pp 1–6

Moriana Group (2004) Service delivery platforms and telecom web services—an industry wide perspective. Moriana Group, Thought leader report, Egham, Surrey, UK

Ohnishi H, Yamato Y, Kaneko M, Moriya T, Hirano M, Sunaga H (2007) Service delivery platform for telecom-enterprise-Internet combined services. In: Global telecommunications conference, 2007. GLOBECOM'07. IEEE, Washington, DC, pp 108–112

Rosenberg J et al (2002) SIP: session initiation protocol. RFC 3261, Internet Engineering Task Force

Shanhe Y, Li C, Li Q (2015) A survey of fog computing: concepts, applications and issues. Workshop on Mobile Big Data. ACM pp 37– 42

Snell J, Tidwell D, Kulchenko P (2009) Programming web services with SOAP. O'Reilly Media, Sebastopol

Sward RE, Boleng J (2012) Service-oriented architecture (SOA) concepts and implementations. In: ACM SIGAda Ada Letters. ACM, New York, pp 3–4

TISPAN (2009) NGN Functional Architecture. ETSI ES 282 001. ETSI, Sophia Antipolis

Waclawsky JG (2005) IMS: a critique of the grand plan. Bus Commun Rev (BCR Enterprises Inc) 35(10):54–58

MIX
Papier aus verantwortungsvollen Quellen
Paper from responsible sources
FSC® C105338

If you have any concerns about our products,
you can contact us on
ProductSafety@springernature.com

In case Publisher is established outside the EU,
the EU authorized representative is:
Springer Nature Customer Service Center GmbH
Europaplatz 3, 69115 Heidelberg, Germany

Printed by Libri Plureos GmbH
in Hamburg, Germany